BIOMAGNETISM AND MAGNETIC BIOSYSTEMS BASED ON MOLECULAR RECOGNITION PROCESSES

To learn more about AIP Conference Proceedings,
including the Conference Proceedings Series, please visit the webpage
http://proceedings.aip.org/proceedings

BIOMAGNETISM AND MAGNETIC BIOSYSTEMS BASED ON MOLECULAR RECOGNITION PROCESSES

Sant Feliu de Guixols, Spain   22 – 27 September 2007

EDITORS

J. Anthony C. Bland

Adrian Ionescu
*University of Cambridge
Cambridge, United Kingdom*

All papers have been peer reviewed.

SPONSORING ORGANIZATIONS

European Science Foundation

European Molecular Biology Organization

Regional Government of Catalonia;
Dept. of Innovation, Universities and
Companies; Comission for Universities and
Research

Melville, New York, 2008
AIP CONFERENCE PROCEEDINGS ∎ 1025

Phys
Sep/oe

Editors

J. Anthony C. Bland

Adrian Ionescu
Cavendish Laboratory
J. J. Thomson Avenue
University of Cambridge
Cambridge CB3 0HE
United Kingdom

L.C. Catalog Card No. 2008928791

ISBN 978-0-7354-0547-9
ISSN 0094-243X

Printed in the United States of America

In memory of

Prof. James Anthony Charles Bland (Tony)

and

Catherine Annietta Bland (Cathy)

CONTENTS

PART 1

MAGNETIC ENTITIES AND MATERIALS FOR
BIOMEDICAL APPLICATIONS

PART 2

MAGNETIC BIOSENSORS AND DETECTION SYSTEMS

Preface

Efficient healthcare is not only a matter of pharmaceuticals, but also a matter of the time required for the diagnosis/analysis of a certain condition or the effective treatment of diseases such as cancer, which ideally should be locally treated with minimal impact to the unaffected parts of the tissue. Furthermore, with the completion of the Human Genome Project, the focus of life sciences has now shifted to the more difficult task of identifying the precise function of each human gene. Therefore, new methodologies and technologies for high throughput gene sequencing and clinical diagnostics are highly sought after. Advances have been made in the recent decade in modern bioassays or medical treatment, however there is still space for improvement related to the speed of detection, cost and specificity.

An alternative approach in this regard can be offered by advances made in the seemingly completely different field of nanomagnetism, in which data processing and its analysis has been accelerated over the last decades by the development of magnetic read/write technology commonly used in electronic industry *e.g.* in hard disks.

By means of combining the expertise of researchers from biology, medicine, biochemistry and physics, the emerging field of "Biomagnetism" aims to apply recent breakthroughs in magnetic nanoscale technologies to biomedical problems, in order to accelerate the development of diagnostic devices and assays. Biomagnetism aims to impact both *in vitro* and *in vivo* biomedicine through the parallel developments of magnetic biosensors and the burgeoning field of "Endomagnetics" where medical researchers are using magnetic particles, injected into the human body for diagnostic or therapeutic purposes. So far only a few interdisciplinary groups around the world, which have recognized the potential afforded in this research, have ventured into this new field of biomagnetism over the last few years. So the time was ripe for a conference in which they could exchange their knowledge, direct the targets of their common research interest and raise awareness in the biotechnology business community.

Therefore as a step to encourage this synergy we have organized the first ever conference dedicated solely to "Biomagnetism and Magnetic Biosystems Based on Molecular Recognition Processes" in Sant Feliu de Guixols, Spain, 22-27 September, 2007, supported financially by the European Science Foundation (ESF) in partnership with the European Molecular Biology Organization (EMBO) and chaired by the late Prof. J.A.C. Bland (University of Cambridge, UK) and co-chaired by Prof. K.R.A. Ziebeck (University of Loughborough, UK). Sixty-two scientists from around the world, from a variety of scientific backgrounds as indicated by the affiliations of the invited speakers, attended the conference. All together they represent most of the groups currently undertaking research in the field of biomagnetism and encompassing all the different aspects of it. The scientific program was divided into eight plenary sessions, each of which included two invited speakers, as well as two poster sessions and concluded with a forward look round table discussion.

The topics covered during the conference were:

- Labelling and Manipulation of Biomolecules
- Applications of Surface Plasmon Resonance
- GMR based Biosensors
- Kinetics of Magnetic Entities in Liquid
- Biomedical Applications of Magnetic Materials
- Microfluidics and Integrated Devices
- Magnetic Nanostructures
- Microfabricated Magnetic Biosensors

While it would be very difficult to include all the topics discussed and the papers presented at the conference in this proceedings issue, we have endeavored to include a comprehensive and diverse selection of them. The 17 articles published here reflect, judging by the amount of attendees at the conference, an enormous activity in this dynamic new research field and presents a compilation of the state-of-the-art biomagnetic technology and therapies. It is therefore our strong belief that this conference proceedings issue will become a timely milestone in the further evolution of biomagnetism.

The Members of the Scientific Committee.

ACKNOWLEDGEMENTS

We are heavily indebted to Ms. Chiara Orefice and the staff at the Hotel Eden Roc, Sant Feliu de Guixols, Spain, for the meticulous logistic organization of the conference. We want also to thank the European Science Foundation, the European Molecular Biology Organization and the Generalitat de Catalunya for their generous financial support. Furthermore, we want to apologize to the contributing authors for the delay of the publication date due to the unforeseen tragic events, which happened during the editing process of this book, and express our gratitude towards the AIP editors, Ms. Kristen Girardi and Ms. Maya Flikop, for their help and understanding. We also acknowledge the help of Prof. K.R.A. Ziebeck for the information supplied regarding Tony's professional career which is printed on the back cover. Finally, A. I. would like to thank all the people in the scientific committee as well as Dr. T. Trypiniotis for refereeing the articles and their help and suggestions regarding the editing. In addition, A. I. would like to acknowledge Prof. J.A.C. Bland and Dr. T. Mitrelias for preparing and submitting the relevant grant proposal to ESF in October 2005. Special thanks also go towards Dr. J. Llandro for numerous advises and discussions regarding the preparation of the conference proceeding issue.

Scientific Committee

- Prof. J.A.C. Bland
- Dr. J. Llandro
- Dr. T.J. Hayward
- Mr. J.J. Palfreyman

- Prof. K.R.A. Ziebeck
- Dr. A. Ionescu
- Dr. T. Mitrelias

Invited Speakers

- Dr. G. Micklem (Department of Genetics, Univ. of Cambridge, UK)
- Prof. U.O. Hafeli (Pharmaceutical Sciences, Univ. of British Columbia, Vancouver, CA)
- Prof. J.A.C. Bland (Cavendish Laboratory, Univ. of Cambridge, UK)
- Dr. H. Brückl (Nano-System-Technologies, Vienna, AT)
- Prof. M.W.J. Prins (Phillips Research and Eindhoven Univ. of Technology, NL)
- Prof. S.X. Wang (Materials Science and Engineering, Stanford Univ., US)
- Prof. J. Bibette (Ecole Supérieure de Physique et de Chemie Industrielles, Paris, FR)
- Prof. Q. Pankhurst (Department of Physics and Astronomy, University College London, UK)
- Prof. K.R.A. Ziebeck (Department of Physics, Loughborough Univ., UK)
- Prof. P.P Freitas (INESC Microsistemas e Nanotecnologias, Lisbon, PT)
- Prof. C.-L. Chien (Department of Physics and Astronomy, John Hopkins Univ., Baltimore, US)
- Prof. A. Sandhu (Tokyo Institute of Technology, JP)

Awards Committee

- Prof. K.R.A. Ziebeck
- Prof. S.X. Wang
- Prof. M.W.J. Prins
- Dr. T. Mitrelias

- Prof. J.A.C. Bland
- Prof. A. Sandhu
- Dr. H. Brückl
- Dr. G. Micklem

Awards

Contributed Talks:
- First Prize: Ms. Kim van Ommering (Eindhoven Univ. of Technology, NL)
- Second Prize: Dr. T.J. Hayward (University of Cambridge, UK)

Contributed Posters:
- Shared First Prize: Ms. S. Bijelovic (Uppsala University, SE) and Dr. J. Llandro (University of Cambridge, UK)

Introduction

As mentioned in the preface, biomagnetism aims to apply recent breakthroughs in magnetic nanoscale technologies to life sciences, in order to advance biomedical research in test-tubes (*in vitro*) and in living systems (*in vivo*) via the parallel developments of endomagnetics and magnetic biosensors based on molecular recognition. It is important first to understand the current practices in biomedicine and molecular biology before the interested reader is really able to appreciate the advantages that biomagnetism can bring.

Over the last thirty years molecular biology has matured as a research field to the extent that it is now possible to rapidly determine the DNA sequences, or genomes, of even complex organisms routinely[1]. This in turn has given rise to the sub-field of "Genomics" in which the behaviour of thousands of genes is examined in single experiments. Key questions that genomics is trying to answer include determining the exact location of all genes in a genome, determining the times and places at which they are switched on and off (or "expressed") and attempting to determine the functions of each gene product. Current estimates suggest the human genome contains 20,000 to 25,000 genes and therefore methods which seek to characterise whole genomes are inherently high-throughput in nature. Other aims of the field include characterising the genetic differences between individuals and looking for correlations between these differences and the physical and medical characteristics of the individuals[2]. It is hoped that such studies will improve the diagnosis and prognosis of disease and in time, through genetic profiling, allow individuals to adapt their lifestyles to minimise the impact of their genetic susceptibility to disease.

Fundamental to molecular biology are a large number of methods for labelling and separating biomolecules which are used in biological assays. Such bioassays aim to identify unknown biomolecules in a sample or to detect the presence of a known subset. The key requirement of a bioassay is to detect binding between probe and target biomolecules, for example an antibody and antigen or the two strands of complementary nucleic acid sequences. One approach to performing such an assay is to use this binding to bring a fluorescent label to the surface of a substrate where it can be measured. The detection of the fluorescent molecules provides quantitative information about the concentration of target molecules present in a solution.

It is notable that the majority of bioassays make use of optical labelling and detection techniques: for instance fluorescence microscopy has been established for many years but more recently the use of intrinsically fluorescent proteins such as the jellyfish green fluorescent protein (GFP) and its spectrally distinct derivatives has

---

[1] D.A. Wheeler *et al.*, *Nature* **452**, 872 (2008)..
[2] Wellcome Trust Case Control Consortium (http://www.wtccc.org.uk).

revolutionised the study of the development of organisms, while the use of fluorescent dyes is a fundamental component of most DNA sequencing and gene expression technologies.

Light-based assays are used both *in vivo* and *in vitro*. The use of the jellyfish GFP *in vivo* has proven to be a very powerful technique: genetic engineering can be used to fuse the GFP gene to any target gene of interest and it has become apparent that in most cases, the resulting target fusion protein retains its normal function. This means that the target's location and behaviour can be examined both at the whole organ or organism level but also with sub-cellular resolution. A limitation of this approach is that only a small number (typically 2-3) of spectrally distinct fluorescent proteins can be simultaneously imaged.

Furthermore, many genomics techniques rely on the sequence-specific recognition properties of nucleic acids such as DNA and RNA: for instance the two anti-parallel strands of the DNA double helix can be melted by increasing the temperature and likewise, if a solution of such single strands is incubated under appropriate conditions the different strands will re-anneal (or "hybridise") in a sequence specific fashion. This forms the basis for many powerful techniques in which probe nucleic acid strands are labelled with fluorescent dyes and used to interrogate biological samples: the labelled molecules seek out and hybridise to complementary sequences present in the sample. In this way extremely detailed information on the location of specific expressed genes in an organism can be obtained. However, the number of probes that can be simultaneously hybridised with a sample is limited to around five, due to the spectral overlap of the dyes used. In some cases, such as in research involving the study of chromosome structural rearrangements it is possible to extend the effective number of dyes. This is done by forming "virtual" dyes from specific ratios of the base set in a technique known as spectral karyotyping[3].

Many biological processes are underpinned by changes in gene expression and another important application of fluorescent dyes in molecular biology is in large-scale gene expression profiling. A gene is said to be expressed when it is copied into an unstable RNA molecule known (for protein-coding genes) as messenger RNA (mRNA). Therefore, an important technique in genomics is the ability to extract mRNA populations from biological samples, label them with dyes and hybridise them to thousands or even millions of distinct probes arrayed on the surface of a glass substrate. After hybridisation, washing and imaging, a high intensity of fluorescence at any given probe spot indicates that the corresponding gene is expressed *i.e.* switched on.

Fluorescent techniques are also central to one of the more sophisticated methods of separating mixtures of cells (or even chromosomes) according to their molecular properties in a process known as flow cytometry[4]. Here, a stream of labelled cells emerges from an oscillating nozzle which breaks the stream up into individual

---

[3] E. Schröck *et al.*, *Science* **273**, 494 (1996).
[4] L.A. Sklar, *Flow Cytometry for Biotechnology*, New York: Oxford University Press, 2005.

droplets. Before this happens there is time to interrogate the stream with laser-light and based on a mixture of fluorescent signals and light scattering a decision can be made to charge the droplet before it breaks from the stream. All droplets subsequently fall between two charged plates, which deflect any charged droplets into collection tubes. In this way pure populations of cells with specific optical properties can be collected at rates of about 20,000 cells per second. As important as the ability of flow cytometry to sort cells is its use as a way of assaying the relative proportions of different labelled cell types within a population. Again, spectral overlap of dyes limits the number of parameters that can be measured simultaneously.

Finally, optical detection plays a key part in all the currently emerging platforms for ultra-high-throughput DNA sequencing[5]. Several of the leading systems use CCD arrays to directly image the addition of differently dyed DNA bases during a step-wise sequencing by synthesis reaction.

The previous examples have shown that the methods commonly used in molecular biology heavily rely on optical techniques and on fluorescent dyes for analytical and detection purposes. Therein, however, lie also the limitations for these techniques such as the spectral overlap of the dyes, the tendency for dyes to be bleached by repeated excitation, the autofluorescence of biological molecules and solutions, and the background fluorescence of the assay supports and detection components.

The question arises whether or not fluorescent dyes can be replaced by alternative labelling methods. Recently, it has become apparent that using magnetic nanoparticles as labels is particularly attractive because such particles can be cheaply produced and are very stable, and hence are unaffected by reagent chemistry or exposure to light. Furthermore, biological samples, being mostly composed of diamagnetic molecules, have an extremely low magnetic background. While it is hard to see how magnetic nanoparticles could replace *in vivo* fluorescent labelling, there is a greater potential for *in vitro* techniques which include examination of thin sections of biological material.

So far the labels used for magnetic biosensing applications have consisted of nanoscopic particles of ferromagnetic metal oxides dispersed within a spherical polymer lattice. These kinds of particles are already widely available with various surface functionalisations as they are already commonly used for separation of mixtures of biological molecules. When subject to an applied magnetic field these particles exhibit a moderate magnetic moment which can be detected by micro-patterned stray field sensors similar to those used in modern hard-disks. A typical sensing device would consist of several arrays of these sensors, each of which is functionalised with a different probe biomolecule. This then allows the device to perform a multiplexed bioassay where the identity of the different probe biomolecules is spatially resolved, as in a conventional fluorescent microarray. Other approaches detect the change in magnetic susceptibility[6] of the whole sample via *e.g.* the change in Brownian relaxation time.

---

[5] E.Y. Chan, *Mutation Research* **573**, 13 (2005).
[6] Magnisense (http://www.magnisense.com).

A variety of magnetic sensors have been applied to the detection of such magnetic labels including anisotropic magnetoresistive sensors, giant magnetoresistive (GMR) sensors, tunnel magnetoresistive sensors, Hall effect sensors, as well as superconducting quantum interference device (SQUID) and fluxgate magnetometers. Most of these types of devices are technologically well established, and hence their optimisation is essentially a matter of operating such sensors within a liquid/biological environment.

Endomagnetics also takes advantage of the properties of small magnetic particles. The researchers and medical doctors working in this area are investigating how magnetic particles, injected into the human body can be used for both diagnostic and therapeutic purposes. Many treatments, such as chemotherapy, currently involve subjecting a patient's entire body to cytotoxic drugs, often resulting in extremely unpleasant side-effects. An alternate approach is the use of nanoparticles to carry a chemical "payload" directly to where it is needed, for example to the site of a tumour (drug targeting). Magnetic particles are particularly attractive for such applications as they can be focused to these locations using external magnetic field gradients.

Another therapeutic application involves chemically targeting magnetic nanoparticles to the site of a tumour and then exciting them with an alternating current (AC) magnetic field. This causes the magnetic moments in the nanoparticles to rapidly oscillate, causing localised heating in the particles and the surrounding tissue. It is hoped that in future this behaviour can be used to directly attack and kill cancerous cells. Furthermore, if magnetic particles can be chemically targeted to the location of a tumour, their presence can be detected using external sensors so as to allow the detection of the position of small tumours without invasive surgery.

Beyond the many potential applications of magnetic based detection techniques to modern biology, there is potential to overcome the limitations on simultaneous use of dyes caused by spectral overlap. For instance coded magnetic particles capable of storing 10 binary digits would in principle generate $2^{10} = 1024$ distinct "dyes", which could be attached to distinct probes allowing the spatial imaging of the expression of 1024 genes simultaneously in tissue slices. Currently such surveys are typically done as one gene at a time[7] and therefore take considerable time and effort. In contrast, the use of coded magnetic particles raises the possibility of being able to screen the gene expression of all genes in a genome in only a few experiments. Likewise, digital encoding of magnetic beads would also allow for the creation of chemical libraries for drug discovery or gene identification.

For portable point of diagnosis tools and "lab-on-chip" devices, magnetic systems show great promise: there is no need for expensive optical detection systems and filters, mass production techniques are already well-established and it is possible to read the signal directly and reproducibly in real time. This may avoid the generation

---

[7] Berkeley Drosophila Genome Project (http://www.fruitfly.org).

and subsequent processing of large images as currently needed for microarray technologies.

Recent work on miniaturised flow-sorting devices using microfluidic channels[8] may well be more compatible with magnetic sensors than conventional flow sorters. At small length scales aqueous fluid flow is laminar and by squeezing the stream it should be possible to bring particles extremely close to detectors while still ensuring a protective layer of fluid is present to cover the sensor.

It has now become apparent that developing a practical magnetic biosensing system is an intrinsically multidisciplinary problem. The magnetic labels used in the assays must have a well defined magnetic moment and size distributions e.g. in order to allow accurate quantification of the number of beads attached to the sensor. Furthermore, it is also desirable in some cases for the magnetic particles to have high magnetic moments to make detection as easy as possible. These are essentially problems of materials science and physics. However, the fabrication of magnetic labels also requires the input of biochemists to ensure the particles have the correct surface functionalisation for a given biological assay in order to minimize non-specific binding of molecules.

Although there is much development work needed to bring any of the above speculative applications to fruition and to optimise the existing biomagnetic technologies, it was apparent from the papers presented at the conference that the time is now ripe for greater effort at this interface between physics, chemistry and biology. A short summary of the conference sessions and the invited speakers' presentations will be useful for the interested reader to deepen their newly acquired knowledge about biomagnetism.

The first session, "Labelling and Manipulation of Biomolecules" was opened by Dr. G. Micklem with a biologist's perspective on the field of biomagnetism, in which the pressing need for faster and more accurate assay techniques was made clear. Highlights included results from Dr. K. Gunnarson (Uppsala Univ.) on precise manipulation of single magnetic beads using magnetised elliptical elements, and from R. Derks (Eindhoven Univ. of Technology) on optimisation of the transport of magnetic carriers through microchannels. In the second session, "Applications of Surface Plasmon Resonance", Dr. H. Brückl highlighted the move from 2D lab-on-a-chip ideas to the 3D lab-on-a-bead concept, which frees magnetic biochips from the limitations of surface-based reaction kinetics. Impressive results using a frequency mixing detection method were presented by Prof. P. Nikitin (Russian Academy of Sciences) who emphasised the importance, both of decreasing the absolute amount of sample needed for detection and the limitation in sensitivity imposed by non-specific binding of molecules. The latter problem of noise due to non-specific binding was raised as an area of concern by several subsequent speakers.

---

[8] Raindance Technologies (http://www.raindancetechnologies.com).

The third session, "GMR-Based Biosensors", was opened by Prof. M.W.J. Prins setting out the vision for a biosensor system as fast (few minutes), cheap and reliable as the diabetes glucose sensor. Prof. S.X. Wang presented assay results from an array of complementary metal–oxide–semiconductor (CMOS)-integrated magnetic sensors, and pointed out the small signals obtained from analogue magnetic sensors as opposed to digitally-switched magnetic random access memory (MRAM) cells. Also notable was research presented by Dr. G. Reekmans (IMEC, Catholic Univ. of Leuven) on optimisation of the binding chemistry and surface charge in sandwich assay-based biosensor applications. Highlights of the fourth session, "Kinetics of Magnetic Entities in Liquid", included a new method of measuring susceptibilities of single beads in solution presented by K. van Ommering (Philips Research/Eindhoven Univ. of Technology), who showed clearly the poor uniformity of commercially available beads. This issue was raised as an area of major concern for magnetic biosensing by several subsequent speakers.

In the fifth session, "Biomedical Applications of Magnetic Materials", both Prof. Q.A. Pankhurst and Dr. A.U.B. Wolter (IFW Dresden) showed how magnetic nanoparticles and Fe-filled carbon nanotubes, respectively, could be used to target drugs into specific cells *in vivo*. Dr. C. Marquina (CSIC-Univ. Zaragoza), speaking on the preparation of core-shell nanoparticles to replace magnetic beads, also demonstrated the necessity for controlling the orientation of biomolecules during attachment to magnetic labels. A very strong sixth session was opened by Prof. P. Freitas (INESC-MN), with his group's integrated handheld single nucleotide polymorphism (SNP) detection platform using thin-film diodes and magnetic tunnel junction sensors. A talk by Dr. C. Liu (IMEC) showed detection of magnetic beads by spin-valve sensors assisted by combined magnetic and dielectrophoretic manipulation, and results from a new kind of digital ring-shaped biosensor were presented by Dr. T. Hayward (Univ. Cambridge). Detection of a few thousand magnetic beads at distances of up to a centimeter was also demonstrated by C. Trigona (Univ. Catania), using a fluxgate-based method.

Notable results in the seventh session, "Magnetic Nanostructures", were presented by Prof. C.-L. Chien on the precise control and almost reversible positioning of nanowires in solution by using AC and DC electrokinetic forces, and also by Dr. B. Ten Haken (Univ. Twente) on characterisation of Fe-filled carbon nanotubes using a micron-sized SQUID "microscope". The final session, "Microfabricated Magnetic Biosensors", was opened by Prof. A. Sandhu, who demonstrated nano-Tesla sensitivity of Hall crosses fabricated in InSb, as well as plans for sensors exploiting the extraordinary magnetoresistance effect in this material.

Finally, during the round table forward look discussion it became apparent that even though GMR and Hall effect sensors are nowadays widely used in the research area, a variety of different sensors such as SQUID or fluxgate magnetometers is still required for specific detection tasks. Furthermore, bioassays, such as those developed and commercialized by Phillips which at the moment can assay four different antigens at a time, are already widely in use in some biochemical laboratories. However, it was pointed out that modern commercially available magnetic beads have been developed for separation and simple manipulation purposes, but their intrinsic problems such as clustering or poor homogeneity of sizes makes them not very well suited for single bead detection purposes. An improvement in homogeneity (magnetic and structural) of the magnetic beads will be crucial for later stages in the development of quicker and more sensitive magnetic bioassays. Furthermore, this new generation of magnetic beads should be ideally only a few nanometers in dimension and provide an improved surface chemistry in order to minimize non-specific binding of organic molecules.

Important questions that need to be addressed include an appraisal of the real commercial applications of magnetic biosensors and in which particular domains such sensors can be of practical use. We hope that this book will help to raise awareness among life scientists to the potential offered by the emerging field of biomagnetism.

The volume is organised into two chapters. The first one is dedicated to magnetic nanoparticles, their fabrication, structural and magnetic properties, motion in liquid, functionalisation and potential uses *e.g.* as digital encoded labels. The second chapter is dedicated to integrated magnetic devices based on different types of sensors for biomedical applications.

Gos Micklem[a], Justin Llandro[b] and Adrian Ionescu[b]

[a] *Department of Genetics, Univ. of Cambridge.*
[b] *Cavendish Laboratory, Univ. of Cambridge.*

Group photo of the conference attendees.

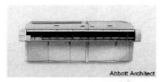
Large robot,
>100 samples per hour

Table-top system,
few samples in parallel

Rapid biosensor

Different sizes of instrumentation: A large robot (top), a table-top system (middle) and a palm-sized biosensor (bottom). Sources: Abbott Diagnostics, BioMerieux and Philips Research. Taken from M.W.J. Prins, **"Magnetic Biosensors – From Molecule to System"**.

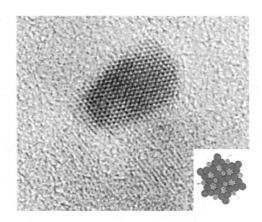

High resolution transmission electron micrograph of a fabricated magnetite superparamagnetic nanoparticle. The lattice structure of magnetite Fe_3O_4 is seen in a [111] projection (schematic representation inset). Taken from N.J. Darton *et al.*, **"The In-flow Capture of Superparamagnetic Nanoparticles for Targeting of Gene Therapeutics"**.

1

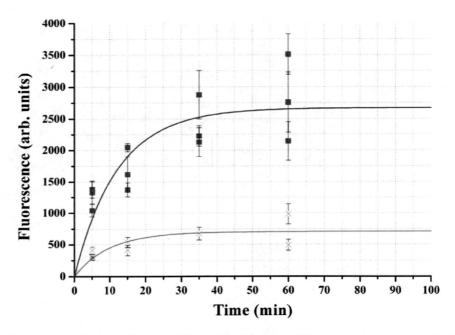

Fluorescent signal vs. time for 12 hybridization experiments with Enterococcus faecium: standard (crosses, red line); with magnetic nanoparticle motion (rectangles, blue line). The lines are guides for the eye. Taken from N. Kataeva *et al.*, **"Progress in Using Magnetic Nanoobjects for Biomedical Diagnostics"**.

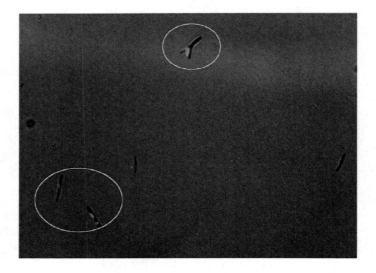

Multisegment nanowires, fluorescent microscope image (40x). The gold layers have been selectively functionalized with fluorescently labeled DNA attached (via a thiol self assembled monolayer). The multisegment character of the nanowires is clear. Taken from F. van Belle *et al.*, **"Templated Growth and Selective Functionalization of Magnetic Nanowires"**.

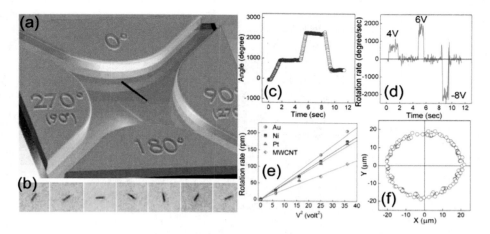

(a) Schematics of rotation of nanowires in suspension by AC voltages applied on the four parts with 90° phase shifts of a quadrupole electrode. (b) Images of rotating Pt nanowires (5.5 μm in length, 0.15 μm in radius) taken every 0.4 sec under 2.5 V at 10 kHz in a 300 μm quadruple electrode. (c) Rotation angle and (d) velocity of Au nanowires (5 μm in length, 0.15 μm in radius) versus time at voltages of 4, 0, 6, 0, -8, 0 V with f = 220 KHz in a 300 μm quadruple electrode. (e) Rotation rate of Au, Ni, and Pt nanowires (10 μm in length, 0.15 μm in radius) and multiwall carbon nanotubes (5 μm in length, 10 nm in radius) versus V^2. (f) The trajectory of a dust particle driven by a bent nanowire as a nano-motor at 10V and 20 kHz. Taken from D. L. Fan *et al.*, "**Controlled Manipulation of Nanoentities in Suspension**".

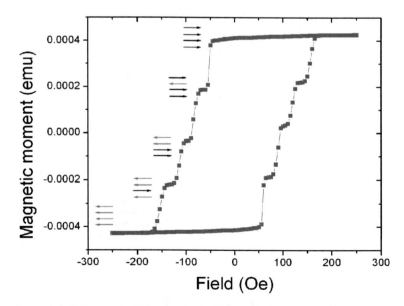

SQUID measurement of four stacked bilayers, Co(3)/PdMn(y) with y = 11, 13, 7 and 9 in order from bottom to top of the sample, each separated by 5 nm of Ta. Each layer switches at a clearly identifiable point. Taken from M. Barbagallo *et al.*, "**Digitally Encoded Exchange Biased Multilayers**".

3

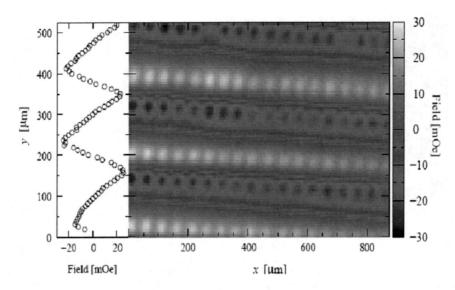

Calibrated magnetic micrograph of the elongated tags. Taken from T. Mitrelias *et al.*, "**Magnetic Microtags and Magnetic Encoding for Applications in Biotechnology**".

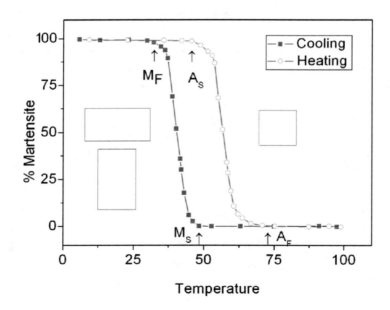

A schematic representation of the austenite (A) - martensite (M) phase transition showing the hysteresis arising from the heating and cooling cycles. The structure of the high symmetry (A) phase is represented by the square and the low-temperature martensitic structure by the rectangles, indicating the presence of variants. Taken from A.P. Gandy *et al.*, "**Magnetically Controlled Shape Memory Behaviour – Materials and Applications**".

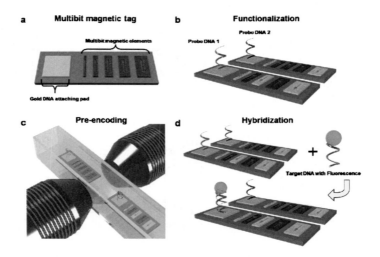

Schematic diagrams for encoding and hybridization of multi-bit digital magnetic tags. (a.) Multibit magnetic tag. (b.) Magnetic tags are functionalized by attaching the probe DNA to the corresponding tag. (c.) Magnetic elements in sequence of aspect ratios are individually addressable and they are pre-encoded depending on the probe DNAs. (d.) Only the tags that show a positive match between the probe and the complementary target DNA will show fluorescence. Taken from B. Hong *et al.*, "**High Throughput Biological Analysis Using Multi-bit Magnetic Digital Planar Tags**".

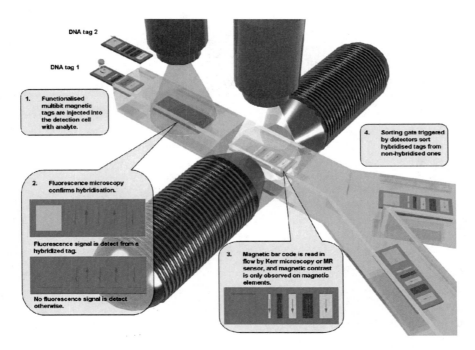

Overview of high-throughput multiplexed biological analysis system. Taken from B. Hong *et al.*, "**High Throughput Biological Analysis Using Multi-bit Magnetic Digital Planar Tags**".

PART 1

MAGNETIC ENTITIES AND MATERIALS FOR BIOMEDICAL APPLICATIONS

Magnetic Biosensors – From Molecule to System

Menno W. J. Prins [a,b]

aPhilips Research, High Tech Campus 4, 5656 AE Eindhoven, NL
bEindhoven University of Technology, P.O. Box 513, 5600 MB Eindhoven, NL

Abstract. Biosensors research combines molecular-level with system-level challenges. The molecular challenge is to rapidly and reliably detect very low concentrations of molecules in a complex biological sample. The system challenge is to find principles that enable the integration of a series of process steps in a disposable cartridge. This paper describes how both aspects can be served by the actuation and detection of magnetic particles, illustrated by the example of immunoassay biosensing.

Keywords: Biosensor, Magnetic particle, Immunoassay.
PACS: 87.85.fk, 87.80.Fe

INTRODUCTION

A biosensor is an instrument that can rapidly measure the concentration of biological molecules in a fluid such as blood or saliva. The most commonly used biosensor is the glucose sensor which is very important for people with diabetes. An example of a commercial glucose sensor is shown in Figure 1. The sensor consists of two components, namely an electronic reader and a disposable cartridge with the shape of a strip. It works as follows. The user inserts the cartridge into the reader instrument. The reader recognizes the presence of the cartridge and asks for a drop of blood. Then the user pricks him- or herself with a tiny needle in order to obtain a small droplet of blood on the skin. The user moves the droplet toward a small opening in the cartridge. When the droplet touches the opening it is drawn into the cartridge and comes into contact with materials inside the cartridge. This initiates a series of biochemical reactions and within a few seconds the instrument indicates what the glucose concentration is in the blood. The user then reads the glucose level and uses this information for an optimal administration of insulin.

The glucose sensor is a very good product: it is small, rapid, easy to use, reliable, and economical (every cartridge costs about 1 Euro). It fits into the everyday life of people with diabetes. You might now wonder why biosensor research is needed, if a high-quality biosensor already exists. The reason is that one would like to measure other substances than just glucose, for example hormones, drugs, proteins, and nucleic acids. These substances have concentrations that are orders of magnitude lower than

CP1025, *Biomagnetism and Magnetic Biosystems Based on Molecular Recognition Processes*
edited by J. A. C. Bland and A. Ionescu
© 2008 American Institute of Physics 978-0-7354-0547-9/08/$23.00

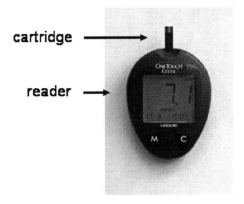

cartridge ⟶

reader ⟶

FIGURE 1. Picture of a commercial glucose sensor. The sensor consists of an electronic reader which operates with a disposable strip-like cartridge.

glucose. So the challenge is to develop biosensors that are much more sensitive, but that are still as small, as rapid, and as reliable as the glucose sensor.

It requires some imagination to realize how small the concentrations are that we want to measure. Let us do a gedankenexperiment. We take two Olympic swimming pools. I blind my eyes while someone throws a single millimeter-sized grain of soluble material in either one of the two pools. We then leave the scene to allow the material to completely dissolve and spread out in the water. When we return we take a sample from each pool and determine the molecular concentration in both droplets. A good sensor is able to determine in which of the two pools the grain was thrown. The corresponding concentration is of the order of a nanogram per liter which is one billion times lower than the concentration of glucose in blood. Such low concentrations can already be detected using large laboratory instrumentation. An aim of biosensors research is to demonstrate the detection of similar and lower concentrations, but now in a compact system, in a sample of just 1 microliter, and in a time of only 1 minute.

Biosensor products are part of the so-called *in-vitro* diagnostics market. This market includes instruments, materials and services that analyze human body fluids in order to evaluate diseases and other medical conditions. Total sales in this market are about 30 billion Euro per year [1]. About 2/3 of the turnover is generated by laboratory testing and about 1/3 by testing outside laboratories. The trend is that more and more testing occurs outside laboratories. Important reasons for this trend are that professionals want to improve the effectiveness and efficiency of their workflows, and that consumers find benefits by testing in their daily lives.

Rapid biosensors can be applied in many domains, e.g. medical, veterinary, food, safety, forensic and environmental applications. Let me give three concrete examples of future applications of biosensors. A first example is in the medical domain. There are a number of low-concentration blood proteins that are indicative of the condition of the heart. Diagnostics and monitoring of heart diseases will improve if medical doctors and individual patients can rapidly and easily measure these proteins. A second example is in pharmaceutical treatment. It is known that patients can respond very differently to pharmaceutical drugs. When a drug is taken at a certain dose, it may work well in some patients while it is ineffective or gives harmful side-effects in other patients. Therefore, it is expected that patients will use biosensors in the future, in order to adjust their medication intake to their personal need. A third example deals with road safety. An important cause of traffic accidents is the use of substances that affect driving ability. Alcohol consumption has long been the biggest problem, but fortunately the number of alcohol related accidents has gone down due to a combination of regular educational campaigns and regular alcohol checks at the roadside. Nowadays, however, the number of fatal accidents caused by the use of illegal drugs is rising. As a consequence there is a great demand for a biosensor suitable for rapid drug testing at the side of the road.

These examples involved the detection of proteins and drugs, two important classes of biological material. There are other classes of materials that are also interesting for rapid-testing applications. One example is the detection of biological cells, for example bacterial cells or white blood cells. Another example is the detection of DNA, the genetic material inside a biological cell. There is a trend of steadily increasing knowledge of how molecules and cells function in the human body due to worldwide bio-medical studies. Such studies regularly generate new markers, i.e. specific molecules that give information about health and disease. New markers will generate new applications for biosensors which is another reason why biosensors research is an important investment for the future.

MOLECULAR BIOSENSORS AND THE SYSTEM CHALLENGE

In this section I want to delve into the technology of molecular biosensors and describe the research challenge from a systems perspective, using protein-based sensors as an example. A topic that requires special attention is the complexity of biological samples.

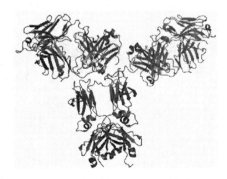

FIGURE 2. Sketch of an antibody, a protein with very strong binding properties. Antibody molecules have a characteristic Y-like shape. The size of the molecule is about 15 nanometers.

Let us assume that we want to detect molecules of type 'A'. In biological samples, the molecules 'A' are surrounded by high and variable concentrations of many other biological molecules. This implies that a biosensor needs to be very specific. We need some kind of selection principle to fish out specifically the molecules 'A' from the complex fluid. Fortunately, nature gives us a helping hand with molecules called antibodies, sketched in Figure 2. Antibodies are large proteins with strong binding properties. The immune system of living beings constantly produces antibodies with the purpose to catch unwanted intruders such as harmful bacteria and viruses. The foundations of antibody-based tests were laid in the 1960's [2]. Antibody technology has significantly progressed, and nowadays one can purchase antibodies against many different molecules. This is very useful for biosensors research, because we can make a biosensor for molecules 'A' by providing a surface with anti-A antibodies. Such a biosensor is called an immuno-sensor. When this surface is exposed to a sample fluid, the molecules in the fluid will bounce against the surface, and molecules of type 'A' will be specifically caught by the anti-A antibodies (see Figure 3).

The surface with antibodies is a central part of the biosensor, but it is not enough. To build a complete biosensor, a sequence of process steps is needed:

- *Sample pretreatment.* A sample normally needs some kind of pretreatment. For example it is passed through a filter, or some (bio-) chemical substances are mixed into the fluid.
- *Transport.* After pretreatment, a transportation process is needed to bring the molecules toward the biosensor surface and into contact with the antibodies.
- *Specific binding.* When the molecules reach the surface, conditions need to be created to enable rapid biological binding between molecules 'A' and the antibodies on the surface. The binding should be specific and non-specific binding to the surface should be minimized.
- *Detection.* Finally, the system should detect how many molecules of type 'A' have been bound to the surface.

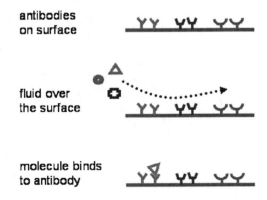

antibodies
on surface

fluid over
the surface

molecule binds
to antibody

FIGURE 3. Top panel: A biosensor surface with antibodies. Middle panel: Fluid flows over the sensor surface. Bottom panel: An antibody captures a molecule from the fluid.

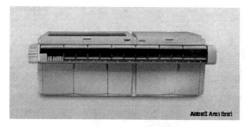

Large robot,
>100 samples per hour

Table-top system,
few samples in parallel

 Rapid biosensor

FIGURE 4. Different sizes of instrumentation: A large robot (top), a table-top system (middle), and a palm-sized biosensor (bottom). Sources: Abbott Diagnostics, BioMerieux, Philips Research.

Now we can define the challenge of biosensors research: the challenge is to build a compact system that integrates all required process steps and that improves the key performance parameters: sensitivity, specificity, speed, accuracy, dynamic range, robustness, ease of use and costs.

Let us take a historical perspective. When antibody-based tests were first performed in the 1960's, a total test involved a long list of manual procedures that needed to be performed in a laboratory. The total test took many hours and errors were often made due to the many manual steps. In the following decades robotized instruments were developed for automated fluid handling and the simultaneous processing of many samples. Such robots are presently used in all clinical laboratories for high-volume blood testing (see Figure 4).

Parallel to the development of large robotized laboratory instrumentation, technologies are being developed for rapid testing outside laboratories. Products presently on the market are mostly based on so-called immunochromatography [2]. The search for improved biosensors has led to the research field called *lab-on-a-chip*, a laboratory-on-a-chip. The word 'chip' refers to a miniaturized system, because computer chip technology is a prime example of miniaturization. In the 1990's the first university groups started to apply miniaturization techniques to the analysis of fluids. In the meantime several research groups have demonstrated that it is in principle possible to perform fluid analysis in a miniaturized device. However, it also became clear that lab-on-a-chip devices often only work properly in a well-controlled environment with well-controlled fluids and operated by well-trained users. For a breakthrough new biosensor technologies are needed which – in addition to being sensitive, specific and rapid – are able to cope with variations in environmental

FIGURE 5. Picture of magnetic particles taken with a scanning electron microscope (SEM). On average the particles have a diameter of 300 nanometer. In our research we study particles in the range between 50 nanometer and 3 micrometer.

conditions, can deal with variabilities in biological samples and are reliable in the hands of unskilled users.

RESEARCH WITH MAGNETIC PARTICLES

Magnetic particles are used in a wide variety of biosensor systems. For example, magnetic particles are used for the extraction of cells from a sample [3, 4]. The particles bind to the cells and the cells are then rapidly extracted by using magnetic fields. The process is robust and rapid, because biological material in itself is hardly magnetic and significant forces can be applied using magnetic fields. Other well-established examples are the use of magnetic particles for nucleic-acid extraction [3, 4] and the use of magnetic particles as carriers in immunoassays [2]. A more recent development is that magnetic particles are used as labels in biological assays, i.e. as indicator tags that allow the quantitative detection of target molecules. The magnetic labels can be detected by a variety of methods, for example by magneto-resistive sensors [5, 6], Hall sensors [7, 8] or optical microscopy [9, 10]. In the following section I will describe a few experiments with magnetic particles and describe an experiment in which magnetic particles are used as labels in a biological assay.

Figure 5 presents an electron-microscopy image of magnetic particles with a diameter of about 300 nanometer. These particles are so-called superparamagnetic particles [3, 4]. They internally contain many ferromagnetic grains with a size of 5-15 nanometers. The grains are so small that they quickly lose their magnetic moment in absence of an external magnetic field. Superparamagnetic particles are readily magnetized to large magnetic moments, yet, the mutual magnetic attraction between different particles can be switched off, preventing irreversible aggregation.

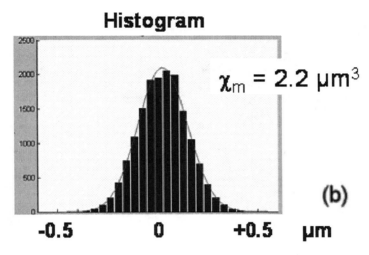

FIGURE 6. Experiment with a single magnetic particle. The panel shows a histogram of particle positions on a chip surface. The histogram reveals the magnetic susceptibility of the particle [11].

Figure 6 shows an optical image of a particle that was suspended in a fluid and placed on a chip surface with a wire [11]. We observed that the particle showed irregular movements caused by the thermal energy. All materials vibrate to some extent due to thermal energy and such vibrations are particularly visible for small objects. In spite of the irregular motion we observed that the particle stayed in the vicinity of the microfabricated wire when the wire was powered with an electrical current. From histograms of observed particle positions (see Figure 6b) we conclude that the confinement of the particle to the wire is caused by the magnetic potential well generated by the electrical current and the magnetism of the particle. The crux of this experiment is that we can derive the susceptibility of an individual nano-particle from the observed irregular motions on a wire [11]. This represents a completely new approach in this field of research.

In another experiment we studied the controlled movement of a magnetic particle on a surface by using multiple wires on a chip [12]. Figure 7a shows the traces of three magnetic particles that hopped between two wires. The wires were alternatingly powered by an electrical current. During the hops of the particles we recorded their positions by a high-speed camera as well as by a magnetic sensor embedded in the chip. The sensor was a giant magneto-resistance (GMR) sensor [5, 13], made from material that strongly changes its electrical resistance when exposed to a magnetic field. Figure 7b shows an experiment in which the number of hopping particles was varied. The data shows that the measured signal scales linearly with the number of hopping particles and that single particles can be detected.

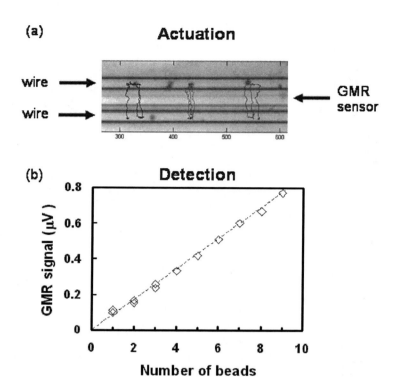

(a) **Actuation**

wire ➜

wire ➜

➜ GMR sensor

(b) **Detection**

GMR signal (µV)

Number of beads

FIGURE 7. Panel (a): Optical microscopy image of a sensor chip with two current wires and a GMR sensor [5, 13]. The traces of three particles are shown, jumping between two wires. The traces were reconstructed from a series of images taken with a high-speed camera. The three particles crossed the GMR sensor due to alternated powering of the wires with an electrical current. Panel (b): The GMR sensor signals recorded as a function of the number of crossing magnetic particles [12].

These experimental demonstrations illustrate the monitoring and control of individual magnetic particles on a chip surface. Such manipulation and detection can enable several functions in a biosensor system:

- *Sample pretreatment.* Magnetic particles can be used to extract biological material from a sample and to agitate and mix fluids [14, 15].
- *Transport.* Magnetic particles can be used as carriers that transport biological molecules toward a chip surface.
- *Specific binding.* Magnetic forces can be applied to the magnetic particles to generate rapid as well as specific biological binding to the chip surface. For example, forces can be applied to concentrate particles at the surface and also to remove weakly bound particles from the surface.
- *Detection.* Magnetic particles can be used as labels, indicating the presence of the molecules of interest on the sensor surface. In addition, we can use the particles to derive functional information of biological molecules, e.g. binding affinities or biological activity, so-called functional biosensing [12].

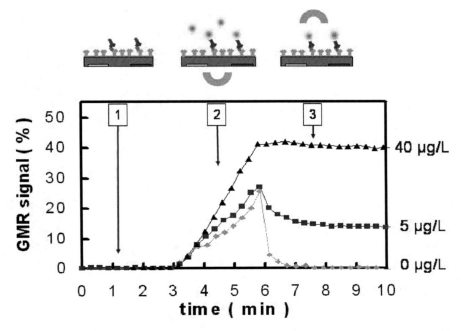

FIGURE 8. Time-trace of a biosensor experiment with magnetic particles. Particles near the chip surface were detected by using GMR sensors embedded in the chip [5, 13]. Three traces are shown from three experiments with different concentrations of parathyroid hormone (PTH), namely 0 μg/L, 5 μg/L, and 40 μg/L. In step 1, PTH molecules were bound to the chip surface and sandwiched between two anti-PTH antibodies. In step 2, magnetic particles were attracted toward the chip surface by magnetic forces and were allowed to bind to the surface. The number of bound magnetic particles became a measure of the number of PTH molecules on the chip surface. In step 3, unbound and weakly bound magnetic particles were pulled away from the chip surface by a magnetic field. The signals clearly depend on the concentration of PTH [16].

The above-mentioned processes all generate challenges and scientific questions. The pretreatment and transportation processes raise questions in the fields of multi-particle dynamics, fluid dynamics, and biological effectiveness. The binding process requires an understanding of the combined physical, chemical and biological forces that appear when a small particle is near to a biologically-activated surface. Finally, the detection process involves questions on biological and physical aspects, concerning signal-over-noise, signal-over-blank and biological function for example.

Figure 8 presents an example of an immuno-biosensor experiment in which several of these processes occur. Three traces are shown from three experiments with different concentrations of a small protein called parathyroid hormone (PTH). The sensor was a silicon chip containing GMR sensors, with a surface coated by anti-PTH antibodies. The graph shows the signal measured by the on-chip GMR sensor as a function of time, indicative of the number of magnetic particles present on the chip surface. The experiment was performed in three steps. In the first step, PTH molecules were bound to the chip surface and subsequently tagged by a second type of anti-PTH antibody.

The result is a sandwich format in which every PTH molecule is sandwiched between two anti-PTH antibodies. In the second step, magnetic particles were attracted toward the chip surface by magnetic forces. These particles had a surface coating with specific binding properties to the second type of anti-PTH antibodies. In this way the number of bound magnetic particles became a measure of the number of PTH molecules on the chip surface. In the third step the unbound and weakly bound magnetic particles were pulled away from the chip surface by a magnetic field. The important point of this series of experiments is that the size of the GMR signal scales with the concentration of PTH, indicating that the device functions as a biosensor.

CONCLUSIONS

I have described how one can control and observe the dynamics of individual magnetic particles on a chip surface and how magnetic particles can be used for biological detection. These examples illustrate the philosophy of the research, namely to use magnetic particles which are large enough to be manipulated and detected by electromagnetic principles, and yet small enough to be substantially affected by biomolecular forces. The biological molecules have a size ranging between 1 nanometer and 20 nanometers and the particles that we study have a size ranging between 50 nanometer and 3 micrometer. The particles can operate in many of the process steps of which a complete biosensor system is composed. Due to the biological compatibility and control possibilities, we believe that magnetic particles will play an ever increasing role in future biosensors, being sensitive, specific, small, rapid and reliable.

ACKNOWLEDGMENTS

I express special thanks for contributions supplied by colleagues at Philips, Eindhoven University, and Future Diagnostics.

REFERENCES

1. Global In Vitro Diagnostics Market Outlook, Report F365 (Frost and Sullivan, 20 May 2005).
2. *The Immunoassay Handbook*, edited by D. Wild, Amsterdam: Elsevier, 2005.
3. Q.A. Pankhurst, J. Conolly, S.K. Jones and J. Dobson, "Applications of magnetic nanoparticles in biomedicine", *J. Phys. D: Appl. Phys.* **36**, R167–R181 (2003).
4. *Scientific and Clinical Applications of Magnetic Carriers*, edited by U. Häfeli, W. Schütt, J. Teller and M. Zborowski, New York: Plenum Press, 1997.
5. M.W.J. Prins and M. Megens, "Magneto-resistive Biosensors", in *Encyclopedia of Materials: Science and Technology*, edited by K.H.J. Buschow, R. Cahn, M.C. Flemings, B. Ilschner, E.J. Kramer, S. Mahajan and P. Veyssiere, Amsterdam: Elsevier, 2007, pp 1-6.
6. M. Megens and M. Prins, "Magnetic biochips: a new option for sensitive diagnostics", *J. Magn. Magn. Mat.* **293**, 702 (2005).
7. H. Fukumoto, K. Takeguchi, M. Nomura and H. Endo, "Rapid and high sensitive bio-sensing system utilizing magnetic beads", in *Proc. 13th Int. Conf. on Solid-State Sensors, Actuators, and Microsystems*, Seoul, Korea, June 5-9 (2005).

8. T. Aytur, J. Foley, M. Anwar, B. Boser, E. Harris and P.R. Beatty, "A novel magnetic bead bioassay platform using a microchip-based sensor for infectious disease diagnosis", *J. Immunol. Methods* **314**, 21-29 (2006).
9. V.N. Morozov and T.Y. Morozova, "Active bead-linked immunoassay on protein microarrays", *Analytica Chim. Acta* **564**, 40-52 (2006).
10. S.P. Mulvaney, C.L. Cole, M.D. Kniller, M. Malito, C.R. Tamanaha, J.C. Rife, M.W. Stanton and L.J. Whitman, "Rapid femtomolar bioassays in complex matrices combining microfluidics and Microelectronics", *Biosensors and Bioelectronics* **23**, 191-200 (2007).
11. K. van Ommering, J.H. Nieuwenhuis, L.J. van IJzendoorn, B. Koopmans and M.W.J. Prins, "Confined Brownian motion of individual magnetic nanoparticles on a chip: Characterization of magnetic susceptibility", *Applied Physics Letters* **89**, 142511 (2006).
12. X.J.A. Janssen, L.J. van IJzendoorn and M.W.J. Prins, "On-chip manipulation and detection of magnetic particles for functional biosensors", *Biosensors and Bioelectronics* **23**, 833 (2008).
13. B.M. de Boer, J.A.H.M. Kahlman, T.P.G.H. Jansen, H. Duric and J. Veen, "An integrated and sensitive detection platform for magneto-resistive biosensors", *Biosensors and Bioelectronics* **22**, 2366 (2007).
14. R.J.S. Derks, A.H. Dietzel, R. Wimberger-Friedl and M.W.J. Prins, "Magnetic bead manipulation in a sub-microliter fluid volume applicable for biosensing", *Microfluidics and Nanofluidics* **3**, 141-149 (2007).
15. F. Fahrni et al., "Polymer flaps with magnetic particles for fluid mixing", to be published.
16. W.U. Dittmer et al., Philips Research, to be published.

The In-flow Capture of Superparamagnetic Nanoparticles for Targeting of Gene Therapeutics

N.J. Darton[a], B. Hallmark[a], X. Han[a], S. Palit[a], M.R. Mackley[a], D. Darling[b], F. Farzaneh[b], and N.K.H. Slater[a]

[a] Department of Chemical Engineering, New Museums Site, Pembroke St, Cambridge. CB2 3RA. UK.
[b] King's College London, Department of Haematological Medicine, The Rayne Institute, 123 Coldharbour Lane, London SE5 9NU, UK.

Abstract. Superparamagnetic magnetite nanoparticles have been synthesised and used for in-flow capture experiments *in vitro* to provide a better understanding of the physical principles that underlie magnetic directed therapy. Experimental observations and modeling work have enabled initial refinement of magnetic targeting strategies and superparamagnetic nanoparticle properties for different therapeutic targeting requirements. It has been discovered that 330 nm and 580 nm agglomerates of 10 nm magnetite cores can be captured with a 0.5 T magnet in flows of up to 0.35 ml•min^{-1} in 410 µm diameter microcapillaries. These flows are typical of blood flow rates found in venules and arterioles in the human cardiovascular system. Further analysis of the data obtained from in-flow capture of superparamagnetic nanoparticles has enabled an initial model to be created, which can be used to estimate the steady state layer thickness of captured superparamagnetic nanoparticles and therefore capillary occlusion at the target site. This work provides the basis for future optimisation of a completely *in vitro* system for testing magnetic directed therapy, enabling data to be provided for preclinical trials.

Keywords: Superparamagnetic nanoparticles, Magnetic capture, In-flow deposition, Targeted therapy.

PACS: 01.30.Cc

INTRODUCTION

Gene therapy is a potentially promising strategy for the treatment of disease that aims to introduce corrective genes to treat conditions which are often difficult to treat with conventional therapeutics. Diseases being investigated for gene therapy treatment include cystic fibrosis, haemophilia A and B, insulin dependent diabetes, AIDS, hepatitis and cancer. One of the most reliable methods of delivering corrective genes to cells in gene therapy utilises the retrovirus as a gene delivery vector. Retroviruses have a low immunogenicity and a wide host range. They can also be modified to be replication defective and so do not harm the host. By using retroviral vectors, highly efficient gene transfer can be achieved resulting in the therapeutic gene being integrated into the genome of the target cell. One of the main difficulties limiting

CP1025, *Biomagnetism and Magnetic Biosystems Based on Molecular Recognition Processes*
edited by J. A. C. Bland and A. Ionescu
© 2008 American Institute of Physics 978-0-7354-0547-9/08/$23.00

effective retroviral gene therapy treatment is the ability to target the retrovirus to the diseased tissue. Hughes *et al.* [1] showed that it was possible to magnetically target retroviral gene therapy vector infection *in vitro* if the retrovirus were attached to a paramagnetic microparticle. A representative result of such a magnetically targeted retroviral infection can be seen in Figure 1. In this demonstration of magnetically targeted gene delivery, a star shape (Figure 1a) was cut from a sheet of 20 mT rubber magnet material. This was attached to the underside of a Petri dish that had an even layer of mammalian tissue (HeLa) cells growing in its base. Paramagnetic microparticle linked retrovirus (Murine Leukemia virus) which had been engineered to deliver a gene conferring resistance to the antibiotic puromycin was then introduced into the Petri dish. Following 30 minutes of agitation the Petri dish was cultured for 24 hrs. The star shaped magnet was then removed and the cells cultured for a further 24 hrs before addition of the antibiotic into the Petri dish for 3 days. The puromycin was toxic to cells that did not receive the puromycin-resistance gene delivered by the retrovirus. The cells that had survived antibiotic treatment were then visualised by staining with Coomassie blue stain. It can be seen from the resulting pattern of surviving cells in Figure 1b that only cells directly over the star shaped magnet were puromycin resistant, therefore demonstrating successful magnetically targeted gene delivery.

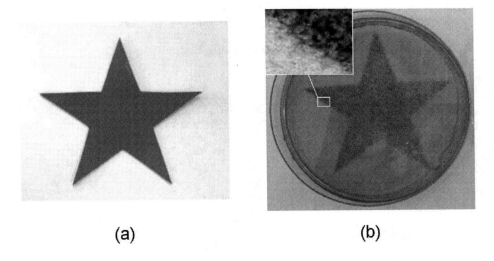

(a)   (b)

FIGURE 1. (a) A five point star cut from a one millimetre thick sheet of rubber magnet material (20 mT) that was used to target gene delivery. (b) Representative result demonstrating magnetically targeted delivery of an antibiotic (puromycin) resistance gene by paramagnetic microparticle linked retrovirus to mammalian tissue cells (HeLa cell line). The precision of magnetic targeting is evident from the sharp boundary between the live antibiotic resistant cells (dark stain) and the clear plate where unresistant cells have not survived and become detached (inset).

Magnetically targeted retroviral gene delivery within the relatively stationary fluid of a Petri dish [1] serves as the first proof of concept for this method; further development would hence allow trials of conditions representative of typical *in vivo* targeting. To date, there have been three published clinical trials of magnetic targeting of chemotherapeutics [2-4], however, the understanding of the complex dynamics involved thus far has been insufficient to allow accurate modeling of different targeting strategies in the human cardiovasculature. The aim of this work was to study magnetically directed targeting of superparamagnetic nanoparticles in microcapillary flows as a step towards the development of an *in vitro* system to more closely mimic *in vivo* flow conditions.

Superparamagnetic Nanoparticle Fabrication

Two synthesis strategies were tested to produce superparamagnetic nanoparticles for magnetic targeting; an organic phase method [5] and an aqueous salt precipitation method [6-7]. The organic phase method [5] involved the high temperature solution phase reaction of iron (III) acetylacetonate with 1,2-hexadecandiol in the presence of oleic acid and oleylamine in toluene. By using this method, the superparamagnetic magnetite particle size could be accurately controlled to be in the 6-12 nm range (Figure 2a) by varying the seed growth time and reaction temperature. Unfortunately, it was found that these nanoparticles formed large agglomerates in the process of transferring from organic to aqueous solvents using surfactants. However, an aqueous solvent is necessary for future clinical applications.

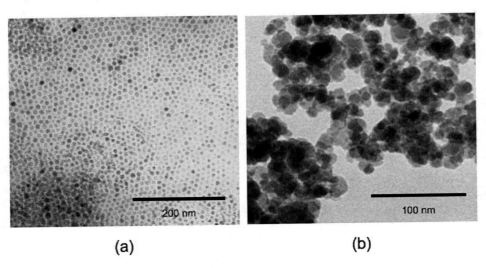

(a)   **(b)**

FIGURE 2. (a) Transmission electron micrograph of 9 nm superparamagnetic nanoparticles fabricated by the organic phase method [5]. (b) Transmission electron micrograph of 10 nm superparamagnetic nanoparticles synthesized by the aqueous precipitation method [7].

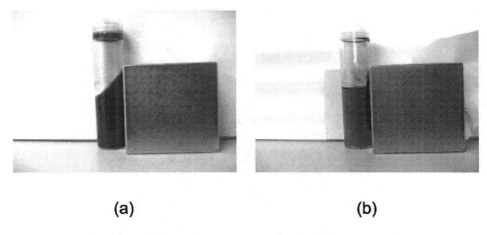

(a)   (b)

FIGURE 3. (a) Glass vial of superparamagnetic nanoparticles in aqueous solution shortly after
positioning next to a 0.5 T NdFeB permanent magnet seen on the right of the vial. (b) The glass vial 5
min. after locating next to the magnet. The superparamagnetic nanoparticles have been pulled down
from solution against the side of the magnet forming a dark layer.

The aqueous precipitation method used solutions of Fe^{2+} and Fe^{3+} chloride; these were co-precipitated with the addition of a base in an oxygen-free, non-oxidizing, environment [7]. This method did not result in such precisely sized nanoparticles as the organic phase method [5], but the particles produced were already soluble in aqueous solutions and were highly superparamagnetic (Figure 3). Zeta-sizing (Brookhaven Zeta-sizer) of the hydrodynamic radii of the nanoparticles produced by the precipitation method indicated that the 10 nm magnetite cores observed in transmission electron microscopy (TEM) (Figure 2b) formed micron sized agglomerates (2.2 μm ±0.1).

Following the method of Mendenhall et al. [8], it was discovered that the agglomerate diameter of superparamagnetic nanoparticles fabricated by the aqueous precipitation method could be well controlled by the addition of 3% w/w polymethacrylic acid (PMAA) to give 575 ± 8.0 nm diameter particles. Subsequent sonication with a 330 W ultrasonic cell crusher (Heat Systems XL-2020) on full power for 10 minutes resulted in 328 ± 3.5 nm diameter agglomerates. High resolution TEM analysis indicated that the 10 nm nanoparticles fabricated had a magnetite Fe_3O_4 lattice structure, as shown in Figure 4. The magnetic properties of the nanoparticles were analysed using SQUID magnetometery at 293 K from -5 T to 5 T. Analysis of the SQUID results showed that the 10 nm particles were superparamagnetic with a magnetic moment of 1.0×10^{-18} A•m^{-1} (in a 628.3 A•m^{-1} field).

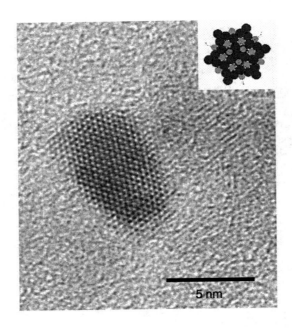

FIGURE 4. High resolution transmission electron micrograph of a fabricated magnetite superparamagnetic nanoparticle. The lattice structure of magnetite Fe_3O_4 is seen in a [111] projection (schematic representation inset). (Dr. Caterina Ducati).

In-flow Capture of Superparamagnetic Nanoparticles

Magnetic targeting of the 580 nm and 330 nm superparamagnetic nanoparticles was investigated by studying their capture in microcapillary flow by a permanent 0.5 T NdFeB magnet [7] (Figure 5). The microcapillary used was part of a novel plastic microcapillary array, termed a microcapillary film or MCF, with 410 µm diameter capillaries. This was made in-house using a novel extrusion process [9, 10]. The MCF was fabricated from Dow® Affinity plastomer which had a low refractive index for optical microscopy. Previous research has characterised fluid flows in these MCFs [11].

A flow of PMAA solution (3% w/w) was passed continuously through a length of the MCF at different constant flow rates ranging between 0.1 ml•min^{-1} to 0.5 ml•min^{-1}. These flow rates corresponded to superficial linear velocities of 1.3 cm•s^{-1} and 6.3 cm•s^{-1}, respectively, in the chosen MCF. The viscosity of the PMAA solution was measured as 2.5×10^{-3} Pa•s and the density as 1024 kg•m^{-3}, giving Reynolds numbers of 1.7 and 8.0, respectively. The viscosity and density of the chosen PMAA solution was comparable to those of blood, with a typical viscosity of circa 3.3×10^{-3} Pa•s [12] and density between 1043 kg•m^{-3} and 1051 kg•m^{-3} [13].

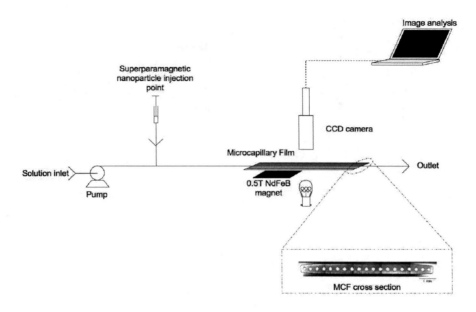

FIGURE 5. Schematic diagram of the apparatus used to capture images of in-flow capture of superparamagnetic nanoparticles in a microcapillary in the microcapillary film.

The 330 nm and 580 nm superparamagnetic nanoparticle agglomerates were introduced at the start of the microcapillary film in a 2 ml pulse of a 40 mg•ml^{-1} particle in 3% (w/w) PMAA suspension. Capture of superparamagnetic nanoparticles in the MCF in the vicinity of the magnet was then recorded by an optical microscope equipped with a CCD camera until either a steady state particle layer was reached for an hour or the layer eroded away. Typically, the lower flow rates resulted in a steady state nanoparticle layer, whereas erosion became the dominant behavior at higher flow rates.

Figure 6 illustrates some typical images recorded during the experiments. From the analysis of these recorded images, it was found that a stable nanoparticle layer formed at superficial linear flow velocities of up to 2.5 cm•s^{-1} for the 580 nm agglomerates and up to 4.4 cm•s^{-1} for the 330 nm agglomerates. These data suggest that smaller nanoparticle agglomerates form a layer that is more impervious to erosion by fluid shear. By balancing the calculated shear stress on the surface of a captured nanoparticle layer with the magnetic force on the superparamagnetic nanoparticles, a model of the system was developed, and used to estimate the steady state thickness of captured nanoparticle layers [14].

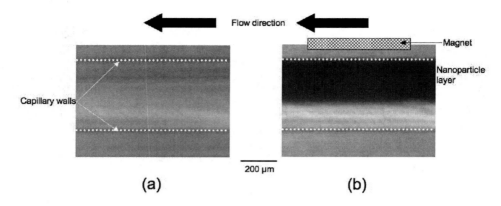

FIGURE 6. (a) Image recorded after superparamagnetic nanoparticles were introduced into a flow of 0.1 ml min^{-1} in absence of the targeting magnet. (b) Image of a stable layer of superparamagnetic nanoparticles captured adjacent to the magnet after their injection into the microcapillary flow. The flow direction is from right to left.

In the constant flow rate regime investigated in this paper, capillary blocking by nanoparticles, analogous to an embolism *in vivo*, was not observed. Current experiments are focusing on the capture of superparamagnetic nanoparticles flowing through an array of 19 microcapillaries as a simplified model of a physiological microcapillary bed. In this arrangement, superparamagnetic nanoparticle capture can be studied in a constant pressure regime that is better representative of cardiovascular blood flow. In the future, magnetically directed therapeutic delivery to a monolayer of cells in the microcapillaries could be examined. This *in vitro* system could then potentially be used in some aspects of preclinical testing.

ACKNOWLEDGMENTS

The authors would like to acknowledge both the BBSRC and EPSRC for funding, Dr. Adrian Ionescu in the Thin Film Magnetism group (Cavendish Laboratory, Cambridge) for his assistance with the SQUID magnetometery measurements and Dr. Caterina Ducati for the high resolution transmission electron microscopy analysis.

REFERENCES

1   C. Hughes, J. Galea-Lauri, F. Farzaneh and D. Darling, *Molecular Therapy* **3** (4), 623-630 (2001).
2   M.W. Wilson, R.K. Kerlan, N.A. Fidelman, A.P. Venook, J.M. LaBerge, J. Koda and R.L. Gordon, *Radiology* **230** (1), 287-293 (2004).
3   A.S. Lübbe, C. Bergemann, H. Riess, F. Schriever, P. Reichardt, K. Possinger, M. Matthias, B. Dorken, F. Herrmann, R. Gürtler, P. Hohenberger, N. Haas, R. Sohr, B. Sander, A.-J. Lemke, D. Ohlendorf, W. Huhnt and D. Huhn., *Cancer Research* **56** (20), 4686-4693 (1996).
4   A.S. Lübbe, C. Bergemann, J. Brock and D.G. McClure, *Journal of Magnetism and Magnetic Materials* **194** (1-3), 149-155 (1999).

5   S.H. Sun, H. Zeng, D.B. Robinson, S. Raoux, P.M. Rice, S.X. Wang and G.X. Li, *Journal of the American Chemical Society* **126** (1), 273-279 (2004).

6   Z.F. Xia, G.B. Wang, K.X.Tao and J.X. Li, *Journal of Magnetism and Magnetic Materials* **293** (1), 182-186 (2005).

7   N.J. Darton, B. Hallmark, S. Palit, X. Han, N.K.H. Slater and M.R. Mackley, *Nanomedicine* **4** (1), 19-29 (2007).

8   G.D. Mendenhall, Y. Geng and J. Hwang, *J Colloid Interface Sci.* **184** (2), 519-26 (1996).

9   B. Hallmark, F. Gadala-Maria and M.R. Mackley, *Journal of Non-newtonian Fluid Mechanics* **128** (2-3), 83-98 (2005).

10   B. Hallmark, M.R. Mackley and F. Gadala-Maria, *Advanced Engineering Materials* **7**, 545-547 (2005).

11   C.H. Hornung, B. Hallmark, R.P. Hesketh and M.R. Mackley, *Journal of Micromechanics and Microengineering* **16**, 434-447 (2006).

12   G. Lowe, A. Rumley, J. Norrie, I. Ford, J. Shepherd, S. Cobbe, P. Macfarlane and C. Packard, *Thrombosis and Haemostasis* **84**, 553-558 (2000).

13   H.G. Hinghofer-Szalkay and J.E. Greenleaf, *Journal of Applied Physiology* **63**, 1003-1007 (1987).

14   B. Hallmark, N.J. Darton, X. Han, S. Palit, M.R. Mackley and N.K.H. Slater, *submitted to Chemical Engineering Science* (2007).

Progress in Using Magnetic Nanoobjects for Biomedical Diagnostics

Nadezhda Kataeva[a], Jörg Schotter[a], Astrit Shoshi[a], Rudolph Heer[a], Moritz Eggeling[a], Ole Bethge[a], Christa Nöhammer[b] and Hubert Brückl[a]

[a] Austrian Research Centers, Division 'Nano-System-Technologies', Donau-City-Str. 1, 1220 Vienna, Austria
[b] Austrian Research Centers, Division 'Life Sciences', 2444 Seibersdorf, Austria

Abstract. A magnetic biochip using the combination of both magnetic nanoobjects as markers and magnetoresistive sensors has proven to be competitive to standard fluorescent DNA-detection at low concentrations. Magnetic nanoobjects additionally provide the unique possibility to actively manipulate biomolecules, on-chip, which paves the way to an integrated 'magnetic lab-on-a-chip' containing detection and manipulation. It is shown that the hybridization process can be accelerated on a biochip. Looking forward, a paradigm change from the 'magnetic lab-on-a-chip' to a 'magnetic lab-on-a-bead' is discussed as a future device solution. The ferromagnetic nanoobjects themselves are thereby directly used both as molecular recognition site and as detection unit.

Keywords: Biochip, GMR, Magnetic Nanoparticle, Diagnostics, Plasmon, Magnetoresistance.

PACS: 61.46.Df, 36.40.Gk, 75.47.De

INTRODUCTION

Recent research progress in the fabrication and characterization of magnetic nanoobjects like rods and beads has triggered many ideas and possible applications, also in the biomedical field. Common applications are e.g. contrast enhancement in imaging, in-vivo drug targeting, cancer treatment by hyperthermia, and labeling on biochips [1, 2]. The size of the nanoobjects ranges from a few nm up to few 100nm and can be reliably reproduced by physical or chemical processes. In the following, such nanoobjects are applied to trace and move biomolecules in fluids for biomedical diagnostics purposes.

MAGNETIC LAB-ON-A-CHIP

The idea of integrating standard laboratory diagnostics into easy-to-use portable devices has received growing attention both by researchers and by biotechnology companies. A recent development is to combine magnetic markers and magnetoresistive sensors in a magnetic biochip. The magnetic nanoparticles and so-called beads which are used as markers are commercially available in a wide range of

CP1025, Biomagnetism and Magnetic Biosystems Based on Molecular Recognition Processes
edited by J. A. C. Bland and A. Ionescu
© 2008 American Institute of Physics 978-0-7354-0547-9/08/$23.00

sizes, functionalities and with a variety of magnetic properties. Such systems promise a number of advantages compared to standard fluorescence biochips. First of all, the magnetoresistive sensors are compatible with the established semiconductor process technology and directly provide an electronic signal suitable for automated analysis. They are scaleable and can be tailored to meet any desired functionality. Furthermore, there is no disturbing background signal like in the case of fluorescent methods. Contrary to fluorescent markers, magnetic markers are stable so that measurements can be repeated many times.

Magnetic markers have proven to have a higher sensitivity at the detection of bio-molecules at low concentrations, as compared to the established fluorescent labeling method [3, 4]. Superparamagnetic microspheres are thereby detected via giant or tunnel magnetoresistance sensors. A further advantage of magnetic nanoobjects is their use as manipulable carriers; manipulable either by external magnetic fields or on-chip via currents running through specially designed line patterns on a chip platform. This is a complementary methodology to dielectrophoresis in which a force is exerted on a dielectric particle when it is subjected to a non-uniform electric field [5]. A combination of dielectrophoretic and magnetic forces could extend the possibilities of particle attraction and repulsion.

REACTION ACCELERATION BY MOVING NANOOBJECTS

By applying magnetic gradient fields, magnetic nanoparticles can be manipulated on-chip, which for example can be utilized to pull the analyte molecules to specific binding sites or to test the binding strength and distinguish between specifically and non-specifically bound molecules [6]. Furthermore, a strong magnetic gradient field can also remove the hybridized analyte DNA and ensure reusability of the biosensor. Finally, if the sensor area decreases for low concentration measurements, it is indispensable that active manipulators accelerate the dwell time of the hybridization step. It can easily be calculated and imagined that it would take years for a single molecule/marker to find a nanoscale sensor only by diffusion.

An acceleration of DNA hybridization can be also achieved by moving magnetic nanoparticles via externally applied fields. Magnetic beads which are immersed in a hybridization solution, e.g. in a fluidic channel, may be moved around and cause a local whirling of the fluid. Figure 1 shows the positive effect on hybridization. Two standard hybridizations were carried out in parallel: in one experiment, superparamagnetic beads were added and actively moved around. The hybridization degree was measured by fluorescence of the target DNA. The magnetic bead supported hybridization was more than three times as effective as the standard procedure in which only diffusive motion is present [7].

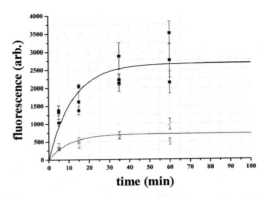

FIGURE 1. Fluorescent signal vs. time for 12 hybridization experiments with Enterococcus faecium: standard (crosses, red line); with magnetic nanoparticle motion (rectangles, blue line). The lines are guides for the eye.

MAGNETIC LAB-ON-A-BEAD

Although a lab-on-a-chip, including the magnetic version, already provides the advantages of a portable and fully automated device, a challenge of future developments is a general device simplification and enhanced sensitivity. Most lab-chips are designed in such a way that surfaces play a major role, either as substrate where molecular reaction takes place or as sensor environment for detection. Proper dealing with substrate surface, microfluidic constrictions, washing and PCR steps for DNA replication make device fabrication and handling complicated and finally unreliable. Therefore, we propose a rugged and easy-to-use solution which would certainly mean a paradigm change from the 'lab-on-a-chip' to a 'lab-on-a-bead' idea. The nanoobjects themselves are thereby directly used both as molecular recognition site and as detector unit via certain changes in its properties; i.e., for example, magnetic relaxation in fluids, precipitation by agglomeration, or plasmon resonance. The magnetic lab-on-a-bead approach promises (a) easier fabrication due to the lack of any chip surface preparation and sensor embedding, (b) easier fluidics as only one reaction and observation chamber is required, (c) and an accelerated reaction time because an intentional motion and tracking of biomolecules are possible.

With the magnetic lab-on-a-bead, two different detection techniques are first choice: magnetorelaxometry and plasmon detection. The magnetorelaxometry measures the Brownian relaxation time of magnetic nanoobjects which depends on their viscous behavior in a fluidic environment [8]. If an externally applied magnetic field which rotates the nanoparticles and orients their magnetization is switched off abruptly, the Brownian rotational motion gives a distinct relaxation profile of the magnetization for biomolecular recognition. The principle relies on the simple fact that enhanced volume and mass due to additionally bound molecules on the surface slows down any motion of the nanoobjects, the Brownian, too. The relaxation time

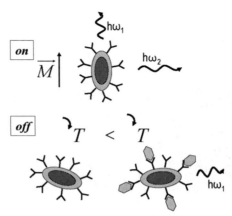

FIGURE 2. Principle of magnetorelaxometry in combination with plasmon detection. 'Loaded' nanoparticles relax slower. This can be detected most sensitively be comparing the two orthogonal plasmon intensities at energies $h\omega_1$ and $h\omega_2$.

depends on whether analyte molecules are bound to the target molecules on the nanoparticles (Figure 2). If the nanoparticles are too small or magnetically isotropic, the relaxation is governed by the thermal Neèl relaxation. If they are too big, they become insensitive to small reaction changes on their shells because their mobility is hardly influenced by the additional 'masses'.

The second possibility is to detect plasmon polaritons in metallic nanoobjects. Plasmons polaritons, i.e. collective excitations of electrons in conductive solids, radiate polarized light at their intrinsic resonance frequency during their decay. The radiation frequency and direction depends on the size and shape of the nanoobjects. Plasmon radiation is sensitive to changes of the refractive index of the surrounding medium, for example any biomolecular binding on the shell. Additional polarizable charges in the medium cause a red shift of the plasmon peak. Numerical modeling shows that core-shell nanoparticles are well suited to such a task [9]. While the magnetic core (e.g. Fe_3O_4) provides externally navigated mobility, Au shells show prominent plasma response. The resonance wavelength can independently be tailored to any point in the visible and infrared spectrum by adjusting the shell thickness. Furthermore, our simulations based on Mie theory reveal that the resolution of any standard spectrometer is sufficient to detect a molecular analyte-target binding process.

Because the plasmonic shifts are still small, we propose a combination of magnetorelaxometry and optical detection with asymmetric, anisotropic, magnetic core-shell nanoparticles for the lab-on-a-bead. Two distinct plasmon modes are available in e.g. elliptical nanoparticles, a low-energetic long axis and a high-energetic short axis mode (Figure 2). The rotation of the nanoparticle due to Brownian motion, for example, switches between these two modes if observed from a certain direction.

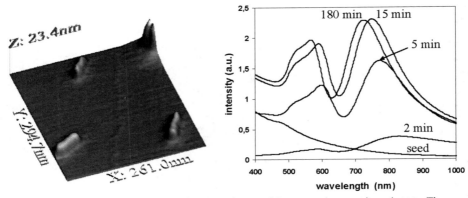

FIGURE 3. Left: Scanning force microscope image of Au nanorods on a mica substrate. The feature in the upper corner represents a bunch of agglomerated nanorods. Right: In-situ optical spectroscopy of Au nanorods shows the development of two plasmon peaks during growth.

Exactly this effect is used to detect the magnetorelaxation time of core-shell nanoparticles, contrary to the usual detection of binding events described in the foregoing paragraph. Thus, the combination of magnetorelaxometry and plasmon excitation offers largest intensity changes instead of a faint peak shift [10].

Even multiplexed detection which is established in many lab-on-a-chip devices by using different sensors in an array to perform different assays can be performed if the information on different binding events is communicated by different colors. This can be achieved by a selected preparation of the beads which assigns different plasmon frequencies to certain functionalities. The plasmon frequency can be adjusted via the shell thickness. Thus, the receptor kind is coded by the plasmon color.

The evolution of a two-plasmon state can be observed in non-spherical Au nanoparticles. Figure 3 shows the optical spectra of Au nanorods during their formation from $HAuCl_4$ precursors in water solution. The Au nanorods were synthesized by employing $AgNO_3$ seeding [11].

The shape of the final nanorods is characterized by scanning force microscopy after spreading on a flat mica substrate (Figure 3, left side). They are typically 10 nm in diameter at a length of 40 to 50 nm. The optical spectrum of the precursor seeding solution is inconspicuous and monotonous. After only two minutes however, two separate peaks already appear in the spectrum around 600 nm and 800 nm wavelengths. These peaks become more pronounced at advanced reaction time, and shift both to blue as the nanorods grow larger. It can be concluded that the Au nanoparticle grow in a rod-like shape from the very beginning of the reaction process, and that the growth continues then in a homogeneous and isotropic mode. The spectrum stagnates after 3 hours indicating the breakdown of the reaction process due to the exhaustion of the precursor material. The nanoparticle state is stable for three month at least since control measurements don't exhibit any changes (not shown). Supposed that the Au nanorods could be aligned, the spatial light radiation and the spectral distribution would be anisotropic according to their orientation. This anisotropy is the base of a huge detection signal if – as mentioned above - magnetorelaxometry and plasmon resonance would be combined at magnetic core-shell nanoobjects.

In conclusion, a paradigm shift from the 'lab-on-a-chip' to a 'lab-on-a-bead' idea is proposed to overcome many of the difficulties connected to the 'lab-on-a-chip' approach. In a 'lab-on-a-bead', the nanoobjects themselves are directly used both as molecular recognition site and as information provider about binding events via characteristic property changes which are remotely detected. We propose the combination of magnetorelaxometry and optical detection of plasmonic excitations. Asymmetric, anisotropic, magnetic core-shell nanoparticles are best suited for this task. However, there is still a tremendous need of research on such nanoparticles in highest quality, i.e. clear core/shell structure, suitable anisotropy of ratio 1:2 or more, functionalized surface for binding biomolecules, high magnetic moment in a non-agglomerating state and narrow size distribution.

ACKNOWLEDGMENTS

The authors J.S. and A.S. acknowledge the financial support of the Austrian FFG (Bridge project, number 810985), and N.K. the stipend from 'Forschung Austria' in the framework of the brainpower Austria program.

REFERENCES

1. D.R. Baselt, G.U. Lee, M. Natesan, S.W. Metzger, P.E. Sheehan and R.J. Colton, *Biosensensors and Bioelectronics* **13**, 731 (1998).
2. Q.A. Pankhurst, J. Connolly, S.K. Jones and J. Dobson, *J. Phys. D: Appl. Phys.* **36**, R167 (2003).
3. J. Schotter, P.B. Kamp, A. Becker, A. Pühler, G. Reiss and H. Brückl, *Biosensors and Bioelectronics* **19**, 1149 (2004).
4. H. Brückl, M. Panhorst, J. Schotter, P.B. Kamp and A. Becker, *IEEE Proc. Nanobiotechnol.* **152**, 41 (2005).
5. J.C. Giddings, *Science* **260**, 1456 (1993).
6. M. Panhorst, P. Kamp, G. Reiss and H. Brückl, *Biosensors and Bioelectronics* **20**, 1685 (2005).
7. R. Heer, M. Eggeling, J. Schotter, C. Nöhammer, R. Pichler, M. Mansfeld and H. Brückl, *J. Magn. Magn. Mater.* **311**, 244 (2007).
8. F. Ludwig, E. Heim, S. Mäuselein, D. Eberbeck and M. Schilling, *J. Magn. Magn. Mater.* **293**, 690 (2005).
9. J. Schotter, O. Bethge and H. Brückl, submitted.
10. Patent pending.
11. B. Nikoobakht and A. El-Sayed, *Chem. Mater.* **15**, 1957 (2003).

Templated Growth and Selective Functionalization of Magnetic Nanowires

F. van Belle[a], J.J. Palfreyman[a], W.S. Lew[b], T. Mitrelias[a] and J.A.C. Bland[a]

[a]Cavendish Laboratory, Cambridge University, J.J. Thomson Avenue, Cambridge, CB3 0HE, UK
[b]School of Physical and Mathematical Sciences, Nanyang Technological University, 1 Nanyang Walk, Singapore 637616

Abstract. Magnetic nanowires are considered as an alternative to magnetic or coloured beads for the labeling of biological entities. Single metal and multisegment magnetic nanowires of lengths between 5-20 μm have been grown, magnetically characterized and released, and selective functionalization of the gold layer with fluorescently labeled DNA is demonstrated. The high magnetic moment of these nanowires, which is scaleable with their length, makes these nanowires a powerful alternative to bead-based labeling techniques.

Keywords: Nanowires, Functionalization, Multisegment.

PACS: 87.85.Ox; 87.85.M-; 87.80.Fe; 85.90.+h; 87.64.-t; 75.75.+a

INTRODUCTION

Biomagnetism has become an area of intense interest in the last few years and nanostructured magnetic materials are beginning to be studied for many applications such as cell manipulation and separation [1], detection of biological compounds in biosensors [2,3], as well as for assisted drug delivery, cancer treatment and other medical and biological applications [4]. Nanowires are a topic of continuing interest in nanotechnology, and there have been many reports on the fabrication and detection of these structures [5]. Because they are very high aspect ratio structures that can be made with a range of different semiconductor or metallic materials, their potential uses are numerous. In addition, magnetic nanowires can easily be manipulated in suspension as well as functionalized with biological or chemical molecules, hence the free flowing magnetic nanowires can be used to manipulate and detect a variety of biological objects such as cells or molecules. The ability to manipulate living cells as well as biomolecules is a great asset in biomolecular research.

Metal or semiconductor nanowires are commonly fabricated by template-assisted electrodeposition. In this method an insulating template is photolithographically defined on a conducting substrate. The template forms the working cathode of an electrolytic cell, where ions from the electrolyte are deposited only onto the developed (conducting) pattern when a driving voltage is applied [6]. Polycarbonate track-etched membranes were amongst the first templates to attract attention [7-10], but form highly irregular arrays. More recently the fabrication of nanowires in home-made or

CP1025, *Biomagnetism and Magnetic Biosystems Based on Molecular Recognition Processes*
edited by J. A. C. Bland and A. Ionescu

commercially available anodized aluminium oxide (AAO) templates has received considerable attention. The anodization of a film of aluminium oxide leads to a highly regular, hexagonally packed array of holes, the depth and diameter of which can be tuned with the duration and voltage of the anodization. After providing a conducting back-contact, metals or semiconductors can be electrodeposited in these holes. The fabrication of noble metal [11–13] and other metal [14–16] nanowires has been investigated in detail by several research groups. Diameters vary from tens to hundreds of nanometers.

Optical detection of these nanowires is extremely difficult, because sub-200 nm diameter is on the limit of optical detection by a normal microscope. However, using highly reflective materials makes it possible to optically detect these nanowires [17–19]. Another possibility is to functionalize the nanowires in such a way that fluorescent molecules can be attached to them, thus enabling the use of fluorescence microscopy [20–22]. Finally, if the nanowires are attached to bigger entities like cells, this allows them to be imaged with optical microscopy [23]. In general, Scanning Electron Microscopy (SEM) is preferred to optical microscopy to image nanowires with diameters up to 200 nm.

Several authors have discussed TEM (transmission electron microscopy), X-ray and magnetometry data on assemblies of different kinds of nanowires, mainly with a sub-100 nm diameter [24–27]. There are several reports on the fabrication of single crystal nanowires [13, 25] and multilayers [28]. The structural characterizations indicate that it is difficult to deposit a pure structure from an electrolyte that contains more than one metal, i.e. if multilayer structures are deposited from a single bath then the different layers are not completely pure.

Semiconductor or conductive nanowires have obvious potential uses in the semiconductor industry, in the form of very small contacts. However, one drawback is the difficulty of manipulating sub-micrometer objects. Nanowires fabricated from metals with different reflective indices have a potential use as very small barcodes, where the traditional black and white stripes have been replaced by non-reflective and reflective materials respectively [17–19]. This can be used as a replacement of current barcodes, but can also be used to label much smaller elements like biological molecules and drugs etc.

Electrodeposited magnetic nanowires are very interesting from a fundamental magnetism point of view, because as structures with a large surface area and high aspect ratio their behavior is significantly different from bulk magnetic materials, and several studies have investigated their magnetization reversal mechanisms [24, 29]. Magnetometry experiments indicate an increase in both coercivity and saturation magnetization with increasing diameter and length. Giant magnetoresistance (GMR) measurements are difficult for these structures due to their small diameter and close packing. GMR measurements have been done on nanowires grown in polycarbonate membranes however, and GMR ratios up to 80% have been measured [7–10]. In a biological environment, detection of labels in general and nanowires in particular can be facilitated by the magnetic content. Whereas they will still be reflective or fluorescent when functionalized or detectable when attached to a bigger entity, their magnetic properties also allows for detection using magnetic sensors [30]. Finally, magnetic nanowires can be manipulated in a fashion unknown to nonmagnetic

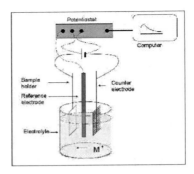

FIGURE 1. Schematic of the electrodeposition setup. .

nanowires. Magnetic field, either static or created by a current, will be able to move these nanowires around; this ability has an enormous influence on their potential uses.

The potential applications of these nanowires are manifold. They can easily be manipulated by a magnetic field [31] and have been shown to align on lithographically defined magnetic patterns [21, 32], thus facilitating incorporation with nanoscale electronics. If magnetic materials are used, they can be detected with a GMR sensor [30], whereas with fluorescent molecules attached they can be detected using a fluorescence microscope [5]. Metal nanowires have special relevance to biological sciences; when magnetic, they can be used to either be incorporated in cells and thus allow cells to be magnetically manipulated [23], or be used as a manipulable support for biological molecules that can be functionalized to the surface [5,20,22]. The fact that multisegment nanowires have an ability to differ from each other by the magnetic segments used, as well as their high magnetic moment, dramatically increases their potential range of applications beyond those of the magnetic beads, the use of which has been explored over the past two decades (e.g. [33]).

In this paper we investigate the magnetic properties of nanowires made from different magnetic materials. We demonstrate that the magnetic moment of a single magnetic nanowire with a diameter of 200 nm and a length of 4 μm can already have a magnetic moment of the order 10^{-10} emu, which scales linearly with the length of the nanowire. The magnetic signal is much larger than that of superparamagnetic beads,

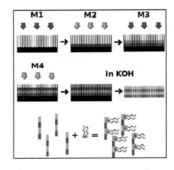

FIGURE 2. Schematic of the growth and functionalization of the multisegment nanowires.

which have a saturation magnetic moment in the order of 1×10^{-11} emu per bead. The release and functionalization of these nanowires is demonstrated, thus opening up a range of potential applications.

Experimental

Nanowires were fabricated by electrodeposition using a setup schematically depicted in Fig.1. Commercially available aluminium oxide porous membranes (Anopore, Whatman) were used as a template. The template is 80 μm thick and the nominal pore size is 200 nm. To create a conducting back-contact, 500 nm copper was sputtered on one side, or if release of the wires was desired, an aluminium layer was inserted between the copper and the AAO template. The back of the substrate was put on adhesive copper tape with the entire back side protected from the electrodeposition solution by a glass microscope slide. The sample was dipped in a home-made electrolyte (0.1 M $NiSO_4$ + 0.1 M boric acid for the Ni nanowires, 0.1 M $CoSO_4$ + 0.1 M boric acid for the Co Nanowires, 0.1 M $NiSO_4$ + 0.1 M $FeSO_4$ + 0.1 M boric acid for the NiFe nanowires), and all deposition was carried out using a potential of -1.0 V (using a Princeton VersaStat II Potentiostat with PowerPulse software). For the gold layer, a commercially available gold solution was used (ECF 60, Metalor). All deposition was done at room temperature with occasional agitation by means of a magnetic stirring bar to improve the homogeneity of the deposition. A platinum mesh was used as a counter electrode and a saturated calomel electrode (SCE) as the reference electrode.

The multisegment wires were grown by sequential electrodeposition whereby the template remains in the cell and the solution is changed after electrodeposition of the first metallic component (schematic in Fig.2). The nanowires length was estimated from the total charge passed, which allowed us to grow nanowires of desired segment lengths. Arrays of multisegment nanowires can thus be produced and dissolution of the template yields a monodisperse suspension of nanostructures that can be functionalized with the desired chemical or biological compounds.

After deposition the sample was cut and a SEM was used for imaging. Magnetic measurements were done with the nanowires in the template using a Superconducting Quantum Interference Device (SQUID). If the nanowires were to be released, the sample was soaked in acetone to release the tape. For the sample with the copper back-contact, the copper was removed using 14% nitric acid. The filled templates are then left to soak overnight in 1 M strong base (KOH) on a hotplate at 80°C. In the case of an aluminium layer inserted between the Cu and the template, treatment with KOH only is sufficient to release the nanowires. The nanowires were then collected by use of a magnet and suspended in either Isopropanol (IPA) or water.

Disulfides have a known strong affinity for gold and form good self assembled monolayers (SAMs) [34]. The compound NHS-C_{10}-S-S-C_{10}-NHS was prepared in-house by reacting 2 equivalents of 11-mercapto-undecanoic acid and NHS (N-HydroxySuccinimide) with 2.4 equivalents of DCC (N,N'-dicyclohexylcarbodiimide) in 99% DMF (dimethylformamide) for 2 hours at room temperature to form the NHS-C_{10}-SH. Air was then bubbled through the solution for 30 minutes to encourage the

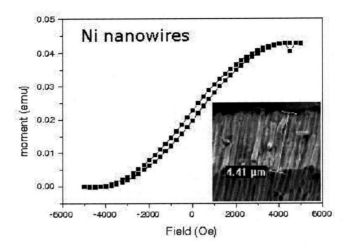

FIGURE 3. SEM picture (inset) and SQUID magnetic measurements of nickel nanowires in their template.

disulfide product, which was then extracted and verified by HPLC (High Performance Liquid Chromatography) and NMR (Nuclear Magnetic Resonance). The SAM was grown on the nanowires by suspension in dry ethanol with 1% disulphide for 2 hours. Excess compound was removed by repeated washing steps with ethanol and then phosphate buffer (PB) (pH 7.5). Then 3 μl of 20 mM dual modified oligo (5'-amino, 3'-FITC) (Sigma-Genosys) were mixed with the nanowires in 1 ml PB in a 1.5 ml eppendorf tube and left overnight to react on an end-over-end rotator. Again excess fluorescent oligo was removed by several cycles of centrifugation and resuspension in fresh PB solvent until the supernatant was no longer fluorescent. For the imaging a Leica DM IRB fluorescence microscope was used, using a drop of the nanowires in solution on a microscope glass slide.

Results

Short magnetic nanowires have been fabricated from nickel, cobalt and permalloy. From the SEM images it is clear the nanowires are around 4 microns long (insets in Fig.3, Fig.4 and Fig.5). The area exposed to the solution was 1.5 cm^2. The nominal pore density is 10^9 pores per cm^2, giving 1.5 × 10^9 pores. The computer was programmed to stop deposition at a charge of 50 Coulombs.

Using Faraday's law, the theoretical thickness for a given area can be calculated with

$$T_{cm} = \frac{M_m Q}{nFA\rho},$$
(1)

where M_m is the molar mass, Q is the deposited charge, n is the valence charge per deposited atom, F is Faraday's constant, A is the surface area in cm^2 and ρ is the density in g/cm^3. Calculating this for the deposited methods gives a theoretical wire length of around 40 μm, rather than the observed 4 μm. The difference will partly be due to hydrogen evolution at the cathode, giving rise to a below 100% current efficiency, and some 'leaking' of charge to the back of the deposition template takes place and a continuous metal film is deposited there before it begins to fill the pores. Finally inhomogeneous wire length, due to variations in the conductivity of the back-contact will also account for some of the discrepancy.

The graphs in Fig.3, Fig.4 and Fig.5 display the magnetic measurements of these nanowires. The field in the SQUID was aligned along the long axis of the wire, background data were subtracted and each measured sample had a surface area of around 6mm^2. As is to be expected, the saturation magnetic moment varies with the magnetic materials, from as high as 0.08 emu for the Co sample to around 0.045 emu for the Ni samples, with the NiFe nanowires having a saturation magnetization in between those values. Despite measuring along the easy axis of the wires, the loops look very hard, needing a magnetic field of almost 5000 Oe in order to reach saturation. This could be due to a strong internal coupling between the nanowires that are very closely packed in the template.

This measured magnetic moment compares favorably with that of the magnetic beads. The magnetic moment of 10 μm spherical superparamagnetic beads (Bangs Lab, BM547) can be calculated to be of the order of 1×10^{-11} emu per bead, whereas from the graphs shown on the short magnetic nanowires it can be seen that a 4 μm long Co nanowire, with a diameter of 200 nm, already has a magnetic moment of the order 10^{-10} emu, which increases linearly with the length of the wire. To give an indication of the different scales of magnetic moment involved: this corresponds to a volume magnetic moment of 1.9×10^{-2} emu/cm^3 for the magnetic beads, compared to 1.6×10^3 emu/cm^3 for the magnetic nanowires.

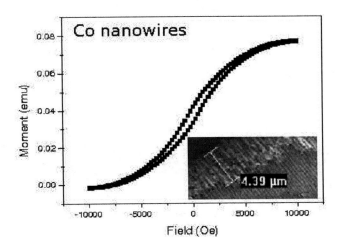

FIGURE 4. SEM picture (inset) and SQUID magnetic measurements of cobalt nanowires in their template.

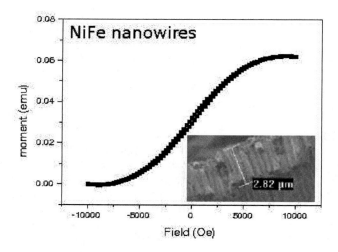

FIGURE 5. SEM picture (inset) and SQUID magnetic measurements of permalloy nanowires in their template.

By alternating the electrolyte a template is submerged in, multisegment nanowires can be fabricated, which are very interesting for the magnetic encoding applications as mentioned in the introduction. In Fig. 6 an SEM image (inset) of long multisegment nanowires can be seen, with three gold layers alternated with two nickel layers. The uneven length of the nanowires is partly due to inevitable damage during sample preparation and partly due to uneven growth.

The magnetic measurements of these multisegment nanowires indicate a saturation moment of 0.02 emu for the sample, which is lower than the single metal nanowires

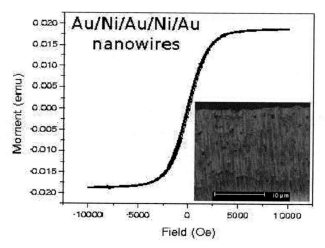

FIGURE 6. SEM picture (inset) and SQUID magnetic measurements of Au/Ni/Au/Ni/Au multisegment nanowires in their template.

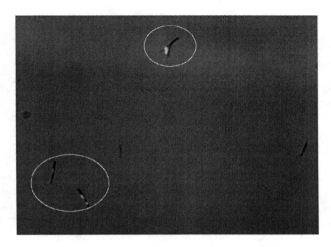

FIGURE 7. Multisegment nanowires, fluorescent microscope image (40x). The gold layers have been selectively functionalized with fluorescently labeled DNA attached (via a thiol self assembled monolayer). The multisegment character of the nanowires is clear.

due to the significant gold sections in the nanowires. Due to these non-magnetic segments the coupling between the nanowires in the template is less, which is reflected in a much lower field of less than 3000 Oe needed to saturate the sample. Although a separation of several microns is used between the two magnetic layers, the magnetic measurements only show one magnetic reversal. This is likely to be due to the still significant coupling of the layers in the template. From the SEM pictures it is clear that the growth of the wires is not homogeneous, which will result in a stronger coupling between the wires.

Dissolving the template and the back-contact releases the nanowires, which can be collected with a magnetic field and re-suspended in solution. The ability to release and manipulate these nanowires is an important asset, opening up possibilities for using them in biology as outlined in the introduction, e.g. cell and molecule manipulation, tracking and detection.

Once the nanowires are in solution, oligonucleotides modified with a fluorescent marker (5'-amino, 3'-FITC, by Sigma-Genosys) can be selectively attached to the gold layers, as demonstrated in Fig. 7. From this image it is clear that the nanowires consist of several gold segments separated by nickel segments. The ability to make multisegment magnetic nanowires that can be detected also by fluorescent microscopy has many potential applications in biology for cell and molecule manipulation and detection [5, 20–23, 32]. By making nanowires with different magnetic segments separated by nonmagnetic segments, nanowires with a different magnetic signature can be created, thus increasing the coding capacity for these types of fluorescent labels.

The ability to both release and functionalize these magnetic nanowires implies that they can be used for the same application as magnetic beads. However, their magnetic characteristics make them more useful than magnetic beads as tags for biological applications. One important drawback of magnetic beads is that they cannot be magnetically encoded. Each type of bead has a fixed magnetic moment and the only possibility for different magnetic codes is to use a different type of bead (of which there are only a handful available). Magnetic nanowires can first of all be made with a large range of magnetic moments by changing the length of the wire. In addition these wires can be encoded with a unique signature by incorporating different magnetic segments, which can either be resolved by binding of fluorescent molecules (much like barcodes) or by distinguishing the magnetic segments with a sensor.

We have demonstrated the fabrication and magnetic measurements of nickel, cobalt and permalloy nanowires, as well as the fabrication, measurement and functionalization with DNA and fluorescent molecules of multisegment magnetic nanowires. It has been demonstrated that the nanowires are good potential candidates for replacing and surpassing the role now taken by magnetic beads. Magnetic nanowires have a much higher magnetic moment: a 4 micron Co nanowire is already an order of magnitude more magnetic than the commonly used superparamagnetic beads, whilst the magnetic moment per volume is five orders of magnitude larger. This facilitates their detection and manipulation. In addition, their magnetic moment is tunable by changing the length of the nanowire, and they form a monodisperse solution.

Currently the length of the nanowires can be fabricated with μm accuracy, but optimizing the electrodeposition setup is expected to bring this down to an accuracy of tens of nm. By combining different magnetic segments in one nanowire, multisegment nanowires can be created. If each of these segments can be detected separately, then a large library of different magnetic (multisegment) nanowires can be generated, which would increase their potential range of applications and give them a significant advantage over magnetic beads. Further research on the biological applications of free flowing functionalized nanowires using an integrated microfluidic chip is in progress [35].

ACKNOWLEDGMENTS

The authors would like to thank J. Dias, M. Lopalco and M. Bradley from Edinburgh University for their expertise and the use of their equipment. Support from the Engineering and Physical Sciences Research Council (EPSRC), UK, is acknowledged.

REFERENCES

1. T. Mitrelias, J.J. Palfreyman, Z. Jiang, J. Llandro, J.A.C. Bland, R.M. Sanchez-Martin and M. Bradley, *Journal of Magnetism and Magnetic Materials* **310**, 2862-2864 (2007).
2. D. Baselt, G. Lee, M. Natesan, S. Metzger, P. Sheehan and R. Coltona, *Biosensors and Bioelectronics* **13**, 731-739 (1998).
3. T. Mitrelias, Z. Jiang, J. Llandro and J.A.C. Bland, *Technical proceedings of the 2006 Nanotechnology conference*, Boston, USA 2006, **2**, 256.
4. J.J Palfreyman, F. van Belle, W.S. Lew, T. Mitrelias and J.A.C. Bland, *IEEE Transactions on Magnetics* **43**, 2439-2441 (2007).

5. D. Reich, M. Tanase, A. Hultgren, L. Bauer, C. Chien and G. Meyer, *Journal of Applied Physics* **93**, 7275-7280 (2003).
6. J. Hulteen and C. Martin, *Journal of Material Chemistry* **7**, 1075-1087 (1997).
7. L. Pireaux, J.M. George, J.F. Despres, C. Leroy, E. Refain, R. Legras, K. Ounadjela and A. Fert, *Applied Physics Letters* **65**, 2482-2486 (1994).
8. S. Dubois, C. Marchal, J.M. Beuken, L. Piraux, J.L. Duvail, A. Fert, J.M. George and J.L. Maurice, *Applied Physics Letters* **70**, 396-399 (1996).
9. A. Blondel, J. Meier, B. Doudin and J.P. Ansermet, *Applied Physics Letters* **65**, 3019-3021 (1994).
10. K. Liu, K. Nagodawithana, P. Searson and C. Chien, *Physical Review B* **51**, 7381-7384 (1995).
11. B. Martin, D.J. Dermody, B.D. Reiss, M. Fang, L.A. Lyon, M.J. Natan and T.E. Mallouk, *Advanced Materials* **11**, 1021-1025 (1999).
12. P. Forrer, F. Schlottig, H. Siegenthalter and M. Textor, *Journal of Applied Electrochemistry* **30**, 533-542 (2000).
13. G. Sauer, G. Brehm, S. Schneider, K. Nielsch, R.B. Whrspohn, J. Choi, H. Hofmeister and U. Gosele, *Journal of Applied Physics* **91**, 3243-3247 (2002).
14. H. He and N. Tao, *Encyclopedia of Nanoscience and Nanotechnology* **X**, 1-18 (2003).
15. A. Yin, J. Li, W. Lian, A. Bennet and J. Xu, *Applied Physics Letters* **79**, 1039-1041 (2001).
16. K. Nielsch, F. Müller, A.P. Li and U. Gösele, *Advanced Materials* **12**, 582-586 (2000).
17. I.D. Walton, S.M. Norton, A. Balasingham, L. He, D.F. Oviso, D. Gupta, P.A. Raju and R.G. Freeman, *Analytical Chemistry* **74**, 2230-2247 (2002).
18. C.D. Keating and M. Natan, *Advanced Materials* **15**, 451-454 (2003).
19. S.R. Nicewarner-Peña, R.G. Freeman, B.D. Reiss, L. He, D.J. Peña, I.D. Walton, R. Cromer, C.D. Keating and M.J. Natan, *Science* **294**, 137-141 (2001).
20. L. Bauer, D. Reich and G. Meyer, *Langmuir* **19**, 7043-7048 (2003).
21. M. Tanase, L.A. Bauer, A. Hultgren, D.M. Silevitch, L. Sun, D.H. Reich, P.C. Searson and G.J. Meyer, *Nano Letters* **1**, 155-158 (2001).
22. A. Salem, P. Searson and K. Leong, *Nature Materials* **2**, 668-671 (2003).
23. M. Tanase, E. Felton, D. Gray, A. Hultgren, C. Chien and D. Reich, *Lab on a Chip* **5**, 598-605 (2005).
24. H. Zeng, R. Skomski, L. Menon, Y. Liu, S. Bandyopadhyay and D.J. Sellmyer, *Physical Review B* **65**, 134426 (2002).
25. M. Aslam, R. Bhobe, N. Alem, S. Donthu and V. Dravid, *Journal of Applied Physics* **98**, 074311 (2005).
26. H. Zeng, M. Zheng, R. Skomski, D.J. Sellmyer, Y. Liu, L. Menon and S. Bandyopadhyay, *Journal of Applied Physics* **83**, 4718-4720 (2000).
27. M. Chen, L. Sun, J. Bonevich, D. Reich, C. Chien and P. Searson, *Applied Physics Letters* **82**, 3310-3312 (2003).
28. R.S. Liu, S.C. Chang, S.F. Hu and C.Y. Huang, *Physica Status Solidi (c)* **3**, 1339-1342 (2006).
29. W. Wernsdorfer, B. Doudin, D. Mailly, K. Hasselbach, A. Benoit, J. Meier, J.-P. Ansermet and B. Barbara, *Physical Review Letters* **77**, 1873-1876 (1996).
30. A. Anguelouch, D. Reich, C. Chien and M. Tondra, *IEEE Transactions on Magnetics* **40**, 2997-2999 (2004).
31. A. Bentley, J. Trethewey, A. Ellis and W. Crone, *Nano Letters* **4**, 487-490 (2004).
32. M. Tanase, D.M. Silevitch, A. Hultgren, L.A. Bauer, P.C. Searson, G.J. Meyer and D.H. Reich, *Journal of Applied Physics* **91**, 8549-8551 (2002).
33. H. Ferreira, D.L. Graham, P.P. Freitas and J.M.S. Cabral, *Journal of Applied Physics* **93**, 7281-7286 (2003).
34. J. Love, L. Estroff, J. Krievel, R. Nuzzo and G.M. Whitesides, *Chemical Reviews* **105**, 1103 (2005).
35. Z. Jiang, J. Llandro, T. Mitrelias and J.A.C. Bland, *Journal of Applied Physics* **99**, 085105 (2006).

43

Controlled Manipulation of Nanoentities in Suspension

D. L. Fan[a, b], R. C. Cammarata[a] and C. L. Chien[a, b]

[a] Department of Materials Science and Engineering, Johns Hopkins University, Baltimore, MD, 21218
[b] Department of Physics and Astronomy, Johns Hopkins University, Baltimore, MD, 21218

Abstract. Nanoentities, a few μm in size, are usually suspended in a liquid to avoid adherence to solid surfaces by the van der Waals forces. Once suspended, however, the nanoentities are in the realm of extremely small Reynolds number (Re) of only 10-5 (Re = 104 for a human swimmer) where viscous drag force overwhelms. Controlled manipulation of nanoentities requires suitable external forces of precise magnitude and direction that can be administered remotely and accurately in a liquid. Here we describe controlled manipulation of nanowires, including both motion and orientation, using dielectrophoretic (DEP) forces via electric potentials applied to patterned electrodes. Nanowires can be compelled to execute controlled linear motion with speeds of 800 μm/sec and rotation motion with rates of 104 rpm, with relevance to NEMS (Nano electromechanical System) devices. Since nanowires, especially multi-segmented nanowires, can be functionalized for chemical and biological purposes, the manipulation of nanowires is also relevant to biomedical applications.

Keywords: Nanoentities, Nanowires, Nanotubes, Manipulation, Rotation, NEMS, MEMS.

PACS: 81.07.-b, 85.85.+j, 47.15.G-, 61.46.-w

INTRODUCTION

Nanoentities, a few μm in size, have become the focus of research in recent years due to their unique attributes. Nanowires are one of them that have high aspect ratios. The diameter ranges from a few nanometers (nm) to a few hundred nm. The length varies between submicrons to tens of microns. The multifunctionalities afforded in multi-segmented nanowires allow tuning of their magnetic [1, 2], electrical [3], and chemical [4] properties for application as nanomemories, nano-generators, and biochemical sensors.

To further exploit their unique properties, nanowires need to be transported and assembled into devices. Since free-standing nanowires readily adhere to solid surfaces due to van der Waals forces, the nanowires are usually suspended in a liquid such as DI water. However, once in the liquid, the nanowires are in the extremely low Reynolds numbers regime of $R_e = av\rho/\eta \approx 10^{-5}$, where a is the size and v is the velocity of the particle, ρ and η are the density and viscosity of the liquid respectively. A low Reynolds number means it is very challenging to manipulate nanowires because

CP1025, *Biomagnetism and Magnetic Biosystems Based on Molecular Recognition Processes*
edited by J. A. C. Bland and A. Ionescu
© 2008 American Institute of Physics 978-0-7354-0547-9/08/$23.00

the viscous force dominates the motion. For example, a nanowire (10 μm in length, 150 nm in radius) moving at a velocity of 100 μm/s in DI water stops within a nm in less than 1 μs after the external driving force has been removed.

Previously, various attempts have been made to manipulate nanowires. Magnetic fields generated from magnets and electromagnets can only align magnetic nanowires [5, 6], because the magnetic force is relatively weak. Optical tweezers can transport individual nanowires using elaborate instruments [7,8]. However, neither magnetic field nor optical tweezers can achieve manipulation on the 10 μm scale. In contrast, controlled manipulation of nanowires, both motion and orientation can be accomplished using electric fields via electric potentials applied to patterned electrodes as described here. Nanowires can be compelled to execute controlled linear motion with speeds of 800 μm/sec and rotation motion with rates of 10^4 rpm. The linear and rotary manipulation of nanowires with high efficiency may be used to assemble nanowires into NEMS devices.

EXPERIMENTAL DETAILS

We used gold (Au), platinum (Pt), and nickel (Ni) as the constituent materials for single-material and multi-segmented nanowires made by electrodeposition through nanoporous templates [1, 2]. The size of the nanopores sets the diameter of the nanowires, whereas the length of the nanowires can be controlled by the deposition time. We have fabricated nanowires with a nominal radius of 150 nm and lengths between 2 μm and 15 μm with aspect ratios between 6:1 and 50:1. The motion of individual nanowires has been captured by an upright optical microscope mounted with a CCD camera operating at 30 frames per second.

RESULTS AND DISCUSSION

Transport of Nanowires in Circular Electrodes

It is known that AC electric field can exert force on small entities in suspension due to the interaction between the induced polarization on the small entities and the applied electric field, an effect known as dielectrophoresis (DEP) [9]. The DEP force can be expressed as [5]:

$$F_{DEP} = p\nabla E = V_{NW}\text{Re}(K_{CM})E\nabla E = \frac{1}{2}V_{NW}\,\text{Re}(K_{CM})\nabla E^2, \qquad (1)$$

where p is the polarization of the particle in the E field, V_{NW} is the volume of the nanoentity, and K_{CM} is the Clausius-Mossotti factor. The current theory of DEP on dielectric materials can be readily extended to metallic nanoparticles. The conducting and geometrical characteristics of the metallic particles as well as the medium in which the particles are embedded affect the DEP force. Our calculation shows that the electric polarization of metallic nanowires with an aspect ratio of 33:1 is enhanced by a factor of 380 comparing with that of spherical metallic spheres [10, 11]. The very low conductivity of the DI water also enhances the DEP effect.

To experimentally characterize the DEP force on Au nanowires in DI water [12, 13], we designed a pair of concentric electrodes composed of an inner and an outer round electrode with radii of 70 μm and 270 μm respectively [Fig. 1(a)]. The advantage of these concentric electrodes is that both the *E* field and the *E* field gradient are in the radial direction with a magnitude proportional to $1/r$ and $1/r^2$ respectively, where r is the distance to the center of the inner electrode. From Eq. (1), we obtain the analytical expression of the DEP force in the gap of the concentric electrodes as

$$F_{DEP} = \frac{\overline{V}^2}{\left(\ln \frac{r_2}{r_1} \right)^2} V_{NW} \varepsilon_m \operatorname{Re}(K_{CM}) \frac{1}{r^3}, \qquad (2)$$

where $\overline{V}$ is the root mean square of voltage, r_1 and r_2 are the radii of inner of outer electrodes respectively, and ε_m is the dielectric constant of the medium. From Eq. (2) the value of DEP force in circular electrodes is expected to be proportional to $1/r^3$. To experimentally determine the DEP force, we suspended the Au nanowires in the gap of the circular electrodes. When the *E* field is applied, all of the randomly suspended nanowires immediately aligned radially, accelerated towards, and finally attached to the inner electrode [Fig. 1(b)]. We captured the motion with the CCD camera and plotted the displacement of nanowires as function of time as shown in Fig. 1(c), from which the velocity (v) and acceleration (a) of nanowires can be obtained by taking the first and second derivative with respect to time. As shown in Fig. 1(d), the velocity of nanowires increases monotonically when approaching the inner electrode, and reaching a speed of 800 μm/sec, which is extremely high for microscopic objects. In comparison, *E-coli* with similar shape and size of nanowires, can only move with a speed of 10 μm/sec.

The equation of motion for a nanowire with mass m, length L_{NW} and radius a_{NW} transported in liquid with viscosity η is governed by

$$ma = F_{DEP} - b\eta v, \qquad (3)$$

where the last term is the drag force due to viscosity, b is the drag coefficient, the value of which can be approximated as $3\pi L_{NW}D$ for nanowires [14]. The value of D depends on the shape of the nanowires. For nanowires 10 μm in length and 0.15 μm in radius, $D = 0.18$. Using the measured velocity and acceleration into Eq. (3), the DEP force can be readily determined. As shown in Fig. 1(e), the acceleration from the DEP force in the gap of the circular electrodes is indeed proportional to $1/r^3$ in agreement with Eq. (2). One also notes that the acceleration due to DEP force is gigantic with a value as high as 0.7 km/sec^2, 70 times the gravitational acceleration. However, the acceleration of nanowires is reduced by 5 orders of

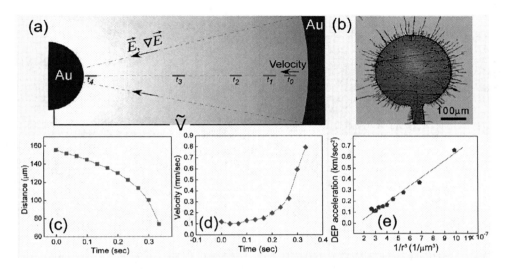

FIGURE 1. (a) Electric field and its gradient are along the radial direction between the Au electrodes. **(b)** Nanowires aligned and attached to the inner electrode. **(c)** Measured displacement and **(d)** velocity versus time of a nanowire at 15 V and 50 MHz. **(e)** DEP acceleration versus $1/r^3$ at 50 MHz, 15 V.

magnitude due to the overwhelming viscous force, as a result of the low Reynolds number of 10^{-5}.

Transport of Nanowires in Quadruple Electrodes

For the circular electrodes described above, the electric field and its gradient are in the same direction. To clarify the different effects of E field and E field gradient in the manipulation of nanowires, we have designed a quadrupole electrode and connected it in a way so that the E field gradient is perpendicular to the E field. As shown in Fig. 2(a), electrodes on the opposite sides of the quadrupole electrode were electrically connected and between them an AC voltage of 10 V at f = 1 MHz was applied. The calculated E field (shown by the lines) and the equipotential curves (shown by the color contours) are shown in Fig. 2(a). Note that the E field gradient is perpendicular to the E field and direct away from the center to the gaps of the electrodes.

We randomly dispensed the Au nanowires in the center of the electrodes (Fig. 3(b)) and recorded the motions of the nanowires. As shown in Fig. 3(b), 3(c), 3(d), and 3(e) taken at 2, 6, 10, and 59 sec respectively after the application of the AC voltage, the nanowires aligned with the E field instantly (Fig. 3(b)), chained (Fig. 3(c)), and transported perpendicular to the orientation towards the high-field regions (Fig. 3(d)-(e)). As a result, the nanowires are depleted from the center and concentrated at the gaps of the electrodes (Fig. 3(e)). The alignment and the assembly of nanowires clearly demonstrate the different roles of the electric field and its gradient. The alignment of the nanowires reveals the actual E field. The transport of the nanowires revealed that the E field gradient determines the transport velocity and direction. By strategically designing microelectrodes, we can configure both the E field and

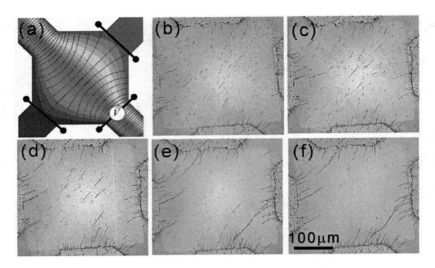

FIGURE 2. (a) Potential (in color contour) and E field (in black lines) distribution in a quadrupole electrode with connected pairs of electrodes. Images of nanowires in suspension at (b) 2 sec, (c) 6 sec, (d) 10 sec, (e) 59 sec, and (f) 180 sec after an AC field of 10V at 1 MHz applied.

the E field gradient in various ways, and thus controlling the orientation as well as the transport of nanowires simultaneously. It should be noted that when the AC voltage is turned off, all of the nanowires remain at their locations, critical for assembling nanowire devices.

Rotation of Nanowires

Next, we describe the rotation of nanowires, a type of manipulation by the AC electric field that only have been exploited recently [15]. We again used the quadrupole electrodes but with four voltages of the same magnitude and frequency but with a sequential phase shift of 90° simultaneously applied to the four electrodes as shown in Fig. 3(a). The gap of the electrodes is 300 μm. These voltages cause the nanowires placed in the central region between the four electrodes to rotate as shown by the snap shots every 0.4 sec for a Pt nanowires 5.5 μm in length with AC voltages of 2.5V at 10 kHz [Fig. 3(b)]. The rotation angle was recorded by analyzing the videos frame by frame as shown in Fig. 3(c). The angle increases with time when clockwise AC voltages were applied, and decreases with counter-clockwise AC voltages. The angle remains unchanged when the voltages were removed. The rotation rate of a nanowire can be determined from the first derivative of the rotation angle with respect of time [Fig. 3(d)]. As a result of the extremely Reynolds number, the nanowires reach the terminal rotary velocity instantly with no apparent acceleration or deceleration, and the rotation instantly stops when the voltages are removed. The rotation rate of a nanowire depends on both the magnitude and the frequency of the applied AC voltage. At a fixed frequency, the rotation rate increases as V^2 as shown in Fig. 3(e). For Au nanowires, the rotation rate increases with frequency from 5 kHz to 100 kHz before gradually decreasing to much smaller values near 1 MHz. In addition to metallic nanowires such as Au, Pt and Ni, we have also rotated multiwall carbon nanotubes (MWCNT), all show the V^2 dependence [Fig. 3 (e)]. The rate of increase appears to scale with the conductivity of the nanoentities, as shown in Fig. 3(e), with Au rotating the fastest and MWCNT the slowest at the same voltage.

The V^2 dependence is advantageous for achieving high rates of rotation. Using a camera with 30 frames/s, we have observed rotation of Au nanowires (10 μm in length, 150 nm in radius) at 10V, with f = 80 kHz in a quadruple electrode with a gap of 150 μm to a speed of 1800 rpm, beyond which our video system could not accurately track the rotation. To explore higher rotation speed, we switched to a high speed CCD camera operating at 2000 frames/sec. We explored different nanowires (Au, Ni, Pt) of various aspect ratios (6.7 to 33), and AC frequencies (5 kHz to 1 MHz). We found that Au nanowires with aspect ratio of about 16 subjected to an AC E field of frequency 80~100 kHz achieved the highest rotation rates. We have achieved for Au nanowires at 20V with f=100 kHz in a quadruple electrode with a gap of 100 μm a record rotation speed of 26,000 rpm, which is still not the limit.

The rotation of nanowires is due to the interaction between the AC E field and the induced electric dipole moment (p), whose value depends on the material and the geometrical shape of the entities, and is proportional to the E field. The rotation of a nanowire is due to the electric torque $T_e = p \times E$, which varies as E^2. For a good conductive nanowire of length L and radius r in a medium of permittivity of ε_m, the electric torque can be calculated as [16]

$$T_e \approx \frac{\pi L^3}{12} \frac{\varepsilon_m}{\ln\frac{L}{r} - 1} E^2$$

(4)

for $L \gg r$. The motion of nanowires rotating in a fluidic medium is governed by $T_e = T_\eta$, where T_η is the viscous torque

$$T_\eta = K\eta\omega$$

(5)

due to the fluidic drag force, where K is the drag coefficient for rotating nanowires, and η is the viscosity of the liquid. Combing equation (4) and (5), we find the terminal angular velocity of nanowires (ω) is proportional to E^2, which accounts for the V^2 dependence observed experimentally.

Another important feature of nanowire rotation by an AC E field is that its chirality can be controlled by the chirality of the rotating E field. The nanowires rotate in the same direction as that of the E field, which is opposite to the phase shift of the AC voltages as shown in Fig. 4(a). The rotation is clockwise (or counter-clockwise) when the phase shift was -90° (or 90°). The chirality of the rotation is determined by the relative charge relaxation times of the nanowire and the medium. When the charge relaxation time of the nanowire is shorter than that of the medium, as in the case of metallic nanowire in DI water, the nanowires rotate in the same chirality as that of the E field, as observed for the frequency range of 5 kHz to 1 MHz.

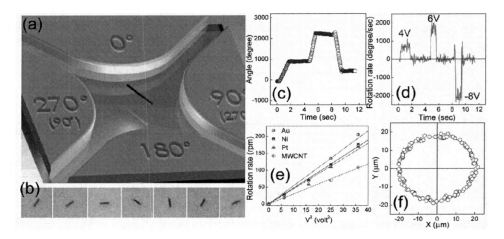

FIGURE 3. (a) Schematics of rotation of nanowires in suspension by AC voltages applied on the four parts with 90° phase shifts of a quadrupole electrode. (b) Images of rotating Pt nanowires (5.5 μm in length, 0.15 μm in radius) taken every 0.4 sec under 2.5 V at 10 kHz in a 300 μm quadruple electrode. (c) Rotation angle and (d) velocity of Au nanowires (5 μm in length, 0.15 μm in radius) versus time at voltages of 4, 0, 6, 0, -8, 0 V with f = 220 KHz in a 300 μm quadruple electrode. (e) Rotation rate of Au, Ni, and Pt nanowires (10 μm in length, 0.15 μm in radius) and multiwall carbon nanotubes (5 μm in length, 10 nm in radius) versus V^2. (f) The trajectory of a dust particle driven by a bent nanowire as a nano-motor at 10V and 20 kHz.

From the above results, we show that the rotation rate, the chirality of rotation, even the total angle of rotation can be precisely controlled. We have achieved a record rotation speed of 26,000 rpm, without reaching the limit yet. To demonstrate the application of nanowire rotation as a NEMS device, we have assembled a nanowire rotor. We used a bent nanowire as the rotor, the chemical connection between the kink of the nanowire and the quartz substrate as the bearing, the quadruple electrodes as the stator, and set it into rotation with an AC voltage of 10V at 20 kHz. A dust particle has been driven into a rotary motion by the two arms of the bent nanowire. Note that the circular trajectory of the dust particle [Fig. 3(f)] occurs only in low Reynolds number regime, where the viscous force overwhelms the inertia of the nanoentities. The dust particle moves only upon impact, and stops as soon as contact ceases.

SUMMARY

In summary, we have demonstrated a highly efficient method for manipulating metallic nanowires in suspension, including transport, acceleration, concentration, collection and rotation by designing AC electric field distribution. Our experiments have clearly shown that nanowires can be precisely transported along prescribed trajectories with controllable orientations and speeds and be rotated to 26000 rpm with controllable rotary rates and chiralities. The manipulation of nanowires may have a wide range of applications for assembling nanowires into NEMS devices, such as sensors, transistors, motors, and stirrers. Since nanowires, especially multi-segmented nanowires, can be functionalized to carry bioentities (e.g. drug, DNA, cell), the manipulation of nanowires can assist transporting these bioentities to the designated places. Our method therefore can be used in many biochemical applications such as gene delivery, cell stimulation, separation of cells and other bioentities, and the study of the mechanical properties of DNA/proteins attached to nanowires.

ACKNOWLEDGMENTS

The work has been supported by NSF DMR04-03849.

REFERENCES

1. T.M. Whitney, J.S. Jiang, P.C. Searson and C.L. Chien, *Science* **261**, 1316-1319 (1993).
2. L. Sun, Y. Hao, C.L. Chien and P.C. Searson, *IBM J. Research and Development* **49**, 79-102 (2005).
3. Z.L. Wang and J.H. Song, *Science* **312**, 242-246 (2006).
4. L.A. Bauer, D.H. Reich and G.J. Meyer, *Langmuir* **19**, 7043-7048 (2003).
5. A. Hultgren, M. Tanase, C.S. Chen and D.H. Reich, *J. Appl. Phys.* **93**, 7554-7556 (2003).
6. C.L. Chien, L. Sun, M. Tanase, L.A. Bauer, A. Hultgren, D.M. Silevitch, G. J. Meyer, P.C. Searson and D.H. Reich, *J. Magnetism and Magnetic Materials* **249**, 146-155 (2002).
7. R. Agarwal, K. Ladavac, Y. Roichman, G. Yu, C.M. Lieber and D.G. Grier, *Opt. Express*, **13**, 8906-8912 (2005).
8. P.J. Pauzauskie, A. Radenovic, E.T. Gnier, H. Shroff, P.D. Yang and J. Liphardt, *Nature Mat.* **5**, 97-101 (2006).
9. H.A. Pohl, *Dielectrophoresis*, Cambridge, Cambridge Univ. Press, 1978.
10. D.L. Fan, F.Q. Zhu, R.C. Cammarata and C.L. Chien, Appl. Phys. Lett. **89**, 223115 (2006).
11. R.D. Miller, *Annual Intern. Conf. of th e IEEE Eng. in Med.and Bio. Soc.*, **12**, 1513-1514 (1990).
12. D.L. Fan, F.Q. Zhu, R.C. Cammarata and C.L. Chien, *Appl. Phys. Lett.* **85**, 4175-4177 (2004).
13. D.L. Fan, F.Q. Zhu, R.C. Cammarata and C.L. Chien, *Appl. Phys. Lett.* **89**, 223115 (2006).
14. J. Happel and H. Brenner, *Low Reynolds number hydrodynamics*, Leyden, Noordhoff international publishing, 1973.
15. D.L. Fan, F.Q. Zhu, R.C. Cammarata and C.L. Chien, , *Phys. Rev. Lett.* **94**, 247208 (2005).
16. T.B. Jones, *Electromechanics of Particles*, Cambridge: Cambridge Univ. Press, 1995.

Digitally Encoded Exchange Biased Multilayers

M. Barbagallo[a], F. van Belle[a], A. Ionescu[a] and J.A.C. Bland[a]

[a]Cavendish Laboratory, University of Cambridge, JJ Thomson Avenue, Cambridge, CB3 0HE, UK.

Abstract. There is a high interest in magnetic multilayers for applications in biotechnology. Exchange biased systems are promising candidates for substituting magnetic beads in more complex applications that require each tag to carry digitally encoded information. We have studied the coercivity and exchange bias of Co/PdMn and CoFe/PdMn samples deposited by dc magnetron sputtering as a function of the materials thickness. We also have fabricated a multilayer comprised of four Co/PdMn bilayers that can be encoded in 16 different states, which are stable at zero applied field, thereby enabling the magnetic tag to store 4 bits of information.

Keywords: Exchange bias, Magnetic biosensing.
PACS: 75.75.+a

INTRODUCTION

Magnetic microparticles or spherical beads have been widely used as labels for antibodies in magnetic immunoassay detection, because of their advantages in assay sensitivity, rapidity, and precision [1-14]. These beads have an advantage over optical encoding mechanisms due to the enhanced detection and manipulation methods possible by utilising their magnetic moment. To date a number of sensitive magnetic field detection devices have been developed for biotechnological applications, these include giant magnetoresistance (GMR), tunneling magnetoresistance (TMR) and anisotropic magnetoresistance (AMR) sensors [15-18]. Microfluidic devices are already being designed that both manipulate and detect these magnetic beads in flow [19]. However, due to their structure, most commonly a polymer matrix with dispersed superparamagnetic particles, only a limited number of different magnetic beads can be made in a certain size range. In order to enhance parallel processing capability and the potential application of magnetic microparticles, it is desirable to have a large variety of uniquely distinguishable magnetic microparticles.

We are currently studying the possibility of substituting the currently used beads with micromagnetic tags, comprised of a stack of distinct ferromagnetic layers, each of which can be individually encoded with one bit of information. This would bring great advantages in parallel processing [20]. The 0 and 1 bits are encoded in the direction of the layer's remanent magnetization vector which is in plane due to the anisotropy induced during growth by applying a magnetic field of 200 Oe. To individually address each bit we have engineered the coercivity and exchange bias (that is the shift of the hysteresis loop) of each layer, by coupling each ferromagnetic

CP1025, Biomagnetism and Magnetic Biosystems Based on Molecular Recognition Processes
edited by J. A. C. Bland and A. Ionescu

layer with an antiferromagnet of varying thickness [21-23]. Exchange bias is an effect first discovered by W.P. Meiklejohn and C.P. Bean [24] in 1956 and is observed in ferromagnetic/antiferromagnetic (FM/AF) systems [25-29], as well as in ferromagnets in contact with their native oxides [30-31]. A review can be found in reference [32]. In order to be able to encode each of the digital states at remanence, we have previously shown that PdMn is an ideal choice, as its exchange bias is very small, while still causing the coercivity of the ferromagnetic layer (Co or CoFe in our samples) to vary over a useful range of 200 Oe [22].

METHODS

The samples were fabricated by dc magnetron sputtering in a Surrey Nanosystems (previously CEVP) chamber with a base pressure between 3 to 9×10^{-9} Torr. The Ar^+ pressure was 3×10^{-3} Torr during the deposition, while a 200 Oe magnetic field was applied in the plane of the sample, which was kept at room temperature and not rotated during the growth. The system is computer controlled and the deposition rate for each material was calibrated by x-ray reflectivity measurements (Figure 1) on a Philips/Panalytical PW3050/65 X'Pert PRO HR horizontal diffractometer. All deposition rates were under 1 nm per second, when deposited at an emission current of 0.1 A. For each material several samples were grown with thicknesses in the range between 2 to 20 nm. The uncertainty in the thickness values quoted below is 5% to 10%. The substrates were scribed from 3" and 5" single crystal p-doped Si(100) wafers, purchased from Si-mat. The 5" wafers had a quoted roughness of 0.1 nm; the roughness from x-ray measurements was consistent rather with a 0.2-0.3 nm value. Ta (5nm) was deposited as a buffer layer, on top of the Si substrate, between each FM/AF bilayer and as capping layer to prevent oxidation and mechanical damage. The Si wafers had a native oxide layer of around 0.1 nm thickness which for brevity is not indicated explicitly in the following sections. The 3" sputter targets were purchased from Testbourne, at 99.9% purity, composite targets had the following at. % composition: $Co_{50}Fe_{50}$, $Pd_{50}Mn_{50}$. In this paper all samples will be indicated by using the example notation Si/Ta(5)/Co(x)/PdMn(y)/Ta(5), where the layer thickness is given in nanometers in the parenthesis. All samples are expected to be polycrystalline.

Measurements of the hysteresis loop {$M(H)$}, coercivity and exchange bias field were performed with a magneto-optical Kerr-effect (MOKE) setup, as well as a Quantum Design MPMS/XL SQUID magnetometer. Two different SQUID magnetometers were used and they were checked for consistency of the results, as well as for the calibration of the MOKE equipment. The focused-spot size of this setup was 3 μm, although the spot was deliberately defocused during the experiments to avoid detecting domain-size effects.

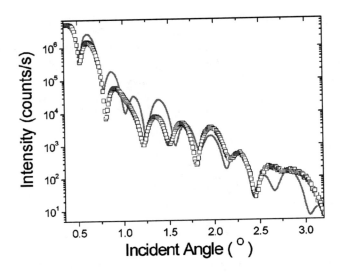

FIGURE 1. X-ray reflectivity measurement (squares) and simulated data (straight line) for a Si/Ta(5)/PdMn(5)/Ta(5) sample.

RESULTS AND DISCUSSIONS

We have measured the coercivity of Co/PdMn bilayers, that is Si/Ta(5)/Co(x)/PdMn(y)/Ta(5), as a function of Co and PdMn thicknesses given as x and y, respectively. A typical hysteresis loop is shown in Figure 2, in the uniaxial in-plane easy axis direction, determined by the direction of the magnetic field applied during growth. As can be seen the switching is quite sharp. We have also measured the $M(H)$ loops at angles of 45° and 90° with respect to the easy axis (not shown). While the coercivity remained constant, the hysteresis loop shape becomes much more slanted (harder) moving away from the easy axis, as expected. The measured exchange bias for the sample in Figure 2 is small, 25 Oe, however the exchange bias data measured by MOKE had a rather large error, around 10 Oe for most of the samples, but up to 30 Oe in some cases. For comparison the multilayer structure shown in Figure 6, measured with a SQUID magnetometer, had an exchange bias between zero and 5 Oe.

In spite of the MOKE imprecision in this regard, it was possible to see a trend in the exchange bias field which remained constant (between 0 and 30 Oe) for increasing PdMn thicknesses of up to 12-13 nm and increased up to and beyond 50 Oe for thicknesses from 14 to 20 nm (not shown). This is consistent with similar data for the CoFe/PdMn system [20]. The MOKE measurements of the coercivity were more precise as it was calibrated against SQUID measurements as well and the uncertainty therefore was estimated to be 5 Oe for all the MOKE measurements.

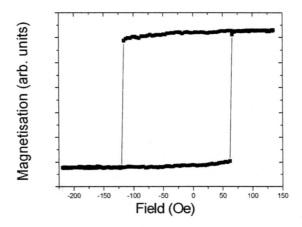

FIGURE 2. *M(H)* MOKE measurement of a Co(3)/PdMn(9) bilayer - Si/Ta(5)/Co(3)/PdMn(9)/Ta(5).

The coercivity measured for different Co/PdMn bilayers is plotted in Figure 3 as a function of PdMn thickness for three different Co thicknesses. The data shows a peak in the coercivity, as expected [22, 32]. In Figure 4 the coercivity is plotted as a function of $1/t$, where t is the Co thickness. There are only three datapoins for each curve but they show the expected $1/t$ dependence [22]. The fits for a PdMn thickness of 6 and 9 nm have an ordinate of 20 Oe in the limit of an infinite Co thickness, as expected since a 300 nm Co film had a measured coercivity of 20 Oe.

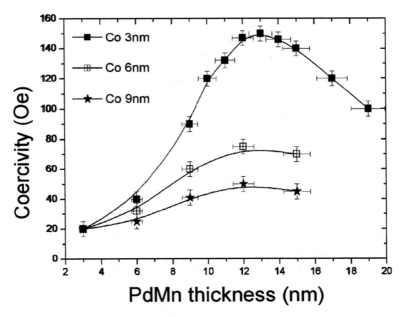

FIGURE 3. Coercivity of Co/PdMn bilayers as a function of PdMn thickness for three different Co thicknesses as measured by MOKE - Si/Ta(5)/Co(x)/PdMn(y)/Ta(5).

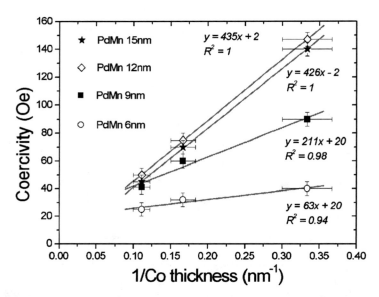

FIGURE 4. Coercivity of Co/PdMn bilayers as a function of (Co thickness)$^{-1}$ for different PdMn thicknesses as measured by MOKE - Si/Ta(5)/Co(x)/PdMn(y)/Ta(5). The fitting equations and R^2 values for each curve is displayed in the same relative order of the fitting curves from top to bottom.

In Figure 5 we present the data for the CoFe/PdMn system. The functional dependence of the coercivity for this system is qualitatively similar to the Co/PdMn system; although the coercivity values in the exchange biased CoFe seem to be higher. The data for the CoFe/PdMn and other exchange biased systems analyzed in the literature is discussed more extensively in reference [22].

Upon stacking two or three multilayers it is still possible to measure the hysteresis loop by MOKE magnetometry, however for four bilayers (due to the limited penetration depth of the He-Ne laser used in the MOKE setup) it is necessary to perform SQUID measurements, as shown in Figure 6. Here, the 3 nm Co layer is biased by four different PdMn thicknesses, which are in order from bottom to top: 11, 13, 7 and 9 nm. Another sample with a different order of the bilayers, that is 7, 9, 11 and 13 nm, showed a similar hysteresis loop, however the two bilayers with a higher coercivity did not exhibit steps as clear as those in Figure 6. This was possibly due to the fact that the roughness increases through the various layers towards the top of the sample, as indicated by x-ray reflectivity measurements, where the simulated fit for the measured data exhibited an increase of about 0.3 nm in the roughness for the top layers, relative to the bottom ones. However, upon depositing the thicker layers (with PdMn thickness of 11 and 13 nm) at the bottom of the sample one can prevent the degradation of the switching sharpness (Figure 6). These two layers have a coercivity value closer to each other than the other two layers. In addition, the increased roughness of the thinner PdMn films seems to have lead to an increased coupling between the FM/AF layers and hence a less sharp switching behaviour.

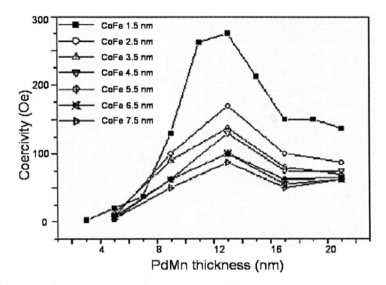

FIGURE 5. Coercivity of CoFe/PdMn bilayers as a function of PdMn thickness for different CoFe thicknesses as measured by MOKE - Si/Ta(5)/CoFe(x)/PdMn(y)/Ta(5). Reproduced from reference [23] and corrected.

The $M(H)$ loop in Figure 6 clearly shows each of the four bilayers switching at a distinct coercivity in agreement with the single bilayer data presented in Figure 3. Moving away from saturation, from the right for example, the Co(3)/PdMn(7) switches first at -50 Oe, while the other three bilayers are still pointing in the opposite direction. The Co(3)/PdMn(9), Co(3)/PdMn(11) and Co(3)/PdMn(13) then switch one at a time upon decreasing the applied magnetic field until the whole sample is at saturation in the opposite direction. The sample was also driven to zero applied field from each of the different states and it was shown to be stable in each encoded configuration when the field was switched off.

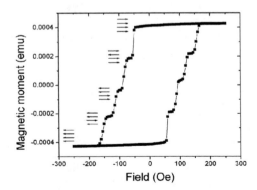

FIGURE 6. SQUID measurement of four stacked bilayers, Co(3)/PdMn(y) with y = 11, 13, 7 and 9 in order from bottom to top of the sample, each separated by 5 nm of Ta. Each layer switches at a clearly identifiable point, at a coercivity in good agreement (within 5 Oe) with the values given in Figure 3.

These results show that it is possible to encode 4 bits of information in such a small multilayer structure, which is $2^4=16$ different states, by identifying each ferromagnetic layer's in-plane magnetization direction as 0 or 1, depending whether it is parallel or antiparallel to the magnetic field direction applied during growth. It is perhaps worth noting that the contribution of each layer to the total sample magnetic moment is constant, which is an additional advantage from the point of view of designing a system to detect such a magnetically encoded multilayer.

CONCLUSIONS

We have systematically investigated the coercivity of Co and CoFe layers exchange biased with PdMn as a function of Co, CoFe and PdMn thickness. All the results here presented are for continuous films, but preliminary work has been carried out on patterned structures (pillar tags) [20] and systematic investigations of the effect of patterning is ongoing in our group.

Furthermore, we have shown that a multilayer comprised of four Co/PdMn bilayers can be encoded in 16 different states which are stable when the applied magnetic field is switched off. Upon substituting the Ta top layer with Au, it is possible to attach biological entities to such magnetic tags. The functionalisation of Au capped structures is detailed in a different paper of these conference proceedings [34]. A microfluidics detection system is also being designed in our group aimed at sensing these magnetic tags in flow.

Finally, micron-sized magnetic multilayer structures could have a variety of interesting applications, most notably as potential information carriers in biotechnology and as a three dimensional extension to conventional magnetic memory [33]. The results here presented confirm the suitability of exchange biased multilayer structures for these purposes.

ACKNOWLEDGMENTS

M. Barbagallo would like to acknowledge the financial support of the Cambridge European Trust and Isaac Newton Trust.

REFERENCES

1. G. Li, V. Joshi, R.L. White, S.X. Wang, J.T. Kemp, C. Webb, R.W. Davis and S. Sun, *J. Appl. Phys.* **93**, 7557-7559 (2003).
2. B.I. Haukanes and C. Kvam, *Bio-Technology* **11**, 60-63 (1993).
3. T. Hultman, S. Stahl, E. Hornes and M. Uhlén, *Nucleic Acids Research* **17** (13), 4937-4946 (1989).
4. M.A.M. Gijs, *Microfluidics and Nanofluidics* **1** (1), 22-40 (2004).
5. D.R. Baselt, G.U. Lee, M. Natesan, S.W. Metzger, P.E. Sheehan and R.J. Colton, *Biosens. Bioelectron.* **13**, 731 (1998).
6. J. Akutsu, Y. Tojo, O. Segawa, K. Obata, M. Okochi, H. Tajima and M. Yohda, *Biotechnol. Bioeng.* **86**, 667 (2004).

7. K. Obata, O. Segawa, M. Yakabe, Y. Ishida, T. Kuroita, K. Ikeda, B. Kawakami, Y. Kawamura, M. Yohda, T. Matsunaga and H. Tajima, *J. Biosci. Bioeng.* **91**, 500 (2001).
8. J.-W. Choi, K.W. Oh, J.H. Thomas, W.R. Heineman, H.B. Halsall, J.H. Nevin, A.J. Helmicki, H.T. Hendersona and C.H. Ahna, *Lab on a Chip* **2**, 27 (2002).
9. R.Wilson, C. Clavering and A. Hutchinson, *J. Electroanal. Chem.* **557**, 109 (2003).
10. G.Li, S.X. Wang and S. Sun, *IEEE Trans. Magn.* **40**, 3000 (2004).
11. H.A. Ferreira, D.L. Graham, P.P. Freitas and J.M.S. Cabral, *J.Appl. Phys.* **93**, 7281 (2003).
12. R.L. Edelstein, C.R. Tamanaha, P.E. Sheehan, M.M. Miller, D.R. Baselt, L.J. Whitman and R.J. Colton, *Biosens. Bioelectron.* **14**, 805 (2000).
13. J. Richardson, P. Hawkins and R. Luxton, *Biosens. Bioelectron.* **16**, 989 (2001).
14. S.P. Mulvaney, H.M.Mattoussi and L.J. Whitman, *Biotechniques* **36**, 602 (2004).
15. J.C. Rife, M.M. Miller, P.E. Sheehan, C.R. Tamanaha, M. Tondra and L.J. Whitman, *Sensors and Actuators A* **107**, 209-218 (2003).
16. M.M. Miller, P.E. Sheehan, R.L. Edelstein, C.R. Tamanaha, L. Zhong, S. Bounnak, L.J. Whitman and R.J. Colton, *J. Magn. Mag. Mat.* **225** (1-2), 138-144 (2001).
17. M.M. Miller, G.A. Prinz, S.F. Cheng and S. Bounnak, *Appl. Phys. Letts.* **81**, 2211 (2002).
18. M.C. Tondra, "Magnetizable Bead Detector", U.S. Patent 2005/0127916A1 (2005).
19. Z. Jiang, J. Llandro, T. Mitrelias and J.A.C. Bland, *J. Appl. Phys.* **99**, 08S105 (2006).
20. F. van Belle, "Magnetic multilayer structures for biological encoding applications", Ph.D. Thesis, University of Cambridge, 2007.
21. F. van Belle, W.S. Lew, T. Mitrelias and J.A C. Bland, *Abstract for the INTERMAG 2006*, San Diego, CA, USA,8-12 May 2006, pp.764.
22. F. van Belle, W.S. Lew, C.A F. Vaz and J.A.C. Bland, *IEEE Trans. Magn.* **42**, 2957-2959 (2006).
23. F. van Belle, T.J. Hayward, J.A.C. Bland and W.S. Lew, *J. Appl. Phys.* **102**, 103908 (2007).
24. W.P. Meiklejohn and C.P. Bean, *Phys. Rev.* **102**, 1413 (1956).
25. H. Xi and R.M. White, *Phys. Rev. B* **61**, 80 (2000).
26. J.P. King, J.N. Chapman, M.F. Gillies and J.C.S. Kools, *J. Phys. D* **34**, 528 (2001).
27. G. Choe and S. Gupta, *Appl. Phys. Letts.* **70**, 1766 (1997).
28. P. Blomqvist, K.M. Krishnan and D.E. McCready, *J. Appl. Phys.* **95**, 8019 (2004).
29. F.Y. Yang and C.L. Chien, *J. Appl. Phys.* **93**, 6829 (2003).
30. J.B. Yi, Z.L. Zhao, J. Ding and B.H. Liu, *J. Appl. Phys.* **97**, 306 (2005).
31. J.B. Yi and J. Ding, *J. Magn. Magn. Mater.* **303**, e160 (2006).
32. J. Nogues and I.K. Schuller, *J. Magn. Magn. Mater.* **192**, 203 (1999).
33. V. Baltz, S. Landis, B. Rodmacq and B. Dieny, *J. Magn. Magn. Mater.* **290–291**, 1286 (2005).
34. F. van Belle, J.J. Palfreyman, W.S. Lew, T. Mitrelias and J.A.C. Bland, elsewhere in this volume.

Magnetic Microtags and Magnetic Encoding for Applications in Biotechnology

Thanos Mitrelias[a+], Theodossis Trypiniotis[a], Frieda van Belle[a], Klaus Peter Kopper[a], Stephan J. Steinmuller[a], J. Anthony C. Bland[a] and Paul A. Robertson[b]

[a]Cavendish Laboratory, University of Cambridge, J.J. Thomson Avenue, Cambridge CB3 0HE
[b]Centre for Advanced Photonics and Electronics, Electrical Engineering Division, Engineering Department, University of Cambridge, 9 J.J. Thomson Avenue, Cambridge CB3 0FA

[+]Corresponding Author: Email: tm10007@cam.ac.uk

Abstract. Microarrays and suspension-based technologies have attracted significant interest recently due to their broad range of applications in medical diagnostics and high throughput molecular biology. The throughput of microarrays is limited by the array density. On the contrary, suspension-based technologies offer a conceptually different approach due to their ability to extend the detection capability by expanding the size of a chemical library (probes). To date, suspension-based methods are limited by the number of distinct codes the solid supports can carry. Here, we present a novel encoding method based on magnetic tags that can be remotely encoded and decoded as they flow in microfluidic channels, by means of magnetic read/write heads. The microtags consist of a substrate and ferromagnetic elements that are individually addressable by the external sensor and can be functionalized with various biochemical probe compounds. We present experimental data demonstrating the reading of individual elements from samples comprising 50×20 μm^2 Ni microbars. The samples are prepared by electrodeposition and read by a micro-fluxgate magnetic sensor set-up for vertical field detection.

Keywords: High throughput bioassays, Magnetic tags, Magnetic biosensors, Micro-fluxgate magnetic sensors.
PACS: 87.85.Ox; 87.85.M-, 87.80.Fe, 85.90.+h, 87.64.-t, 75.75.+a

INTRODUCTION

The design of devices that can be used for high throughput biological screening where samples are simultaneously screened for multiple analytes, has been of interest recently, mainly due to the broad range of exciting biological and biomedical applications, ranging from genome sequencing and genetic analysis to clinical diagnostics. Drug discovery where typical assays involve screening of vast chemical libraries for specific target molecules could also benefit from the development of high throughput methods. These assays are performed by using thousands of chemical reactions; multiplexing is essential for speeding up the process by allowing the simultaneous performance of multiple discrete assays using the same microvolume

CP1025, Biomagnetism and Magnetic Biosystems Based on Molecular Recognition Processes
edited by J. A. C. Bland and A. Ionescu
© 2008 American Institute of Physics 978-0-7354-0547-9/08/$23.00

sample at low cost and with high efficiency. Several challenges need to be addressed for the effective implementation of high throughput technologies though, the most important perhaps is the need to track each reaction and monitor which library compound produced the specific reaction; this is a complex requirement since the number of separate chemical reactions involved is very large. Other challenges include the necessity to maintain the specificity and sensitivity of the assay and assure the reproducibility of the measurements. Two main methods have been developed for multiplex analysis; one is based on arrays of molecular probes where the probe identity is determined from positional encoding (microarrays) and the other is based on probes attached to solid supports in suspension, where the probe identity is derived from encoding each and every support in the solution (suspension-based methods).

Microarrays

The state-of-the-art technology commonly used for performing multiplexed assays, utilises microarrays, which comprise a planar substrate with discrete micron-sized spots consisting of compounds of known molecular identity [1]. This ensemble of spots forms an addressable two-dimensional array of several hundred thousand probes that are used in a variety of biological assays, from gene sequencing, gene expression [2, 3] or single nucleotide polymorphism (SNP) analysis, to drug development and proteomics studies [4, 5]. For example, a DNA microarray, also known as gene chip, has labelled probes consisting of single stranded oligonucleotides of a specific length, which can then interact and hybridise with the target DNA only if their complementary sequences are present in the target solution. The target DNA strands are marked with fluorescent labels and the array is optically scanned to identify the hybridized spots. The target compounds are identified by the spatial coordinates of the location of the microarray spot.

Several challenges are associated with microarray assays, such as the collection of reliable and reproducible data in one experiment and the necessity for very rigid user protocols and extensive user expertise, while extraneous factors may also influence the measurements [6, 7]. Moreover, they suffer from a fundamental limitation due to the nature of the method with which information is encoded, i.e. by spatially resolved signals. The density of the inherently 2D positional data that is used to identify a particular spot can not increase substantially due to geometric constraints, including the limitation of cross-talk from neighbouring spots. In addition, an ultimate limitation on the spatial resolution is introduced by the optical wavelength used to interrogate the array. The speed of the assay is restricted by the slow diffusion of the target molecules to the surface of the array and by limitations in the range of analyte concentrations that can be simultaneously detected [8].

Suspension-Based Technologies

Conceptually different high throughput technologies are being developed to address these challenges. Suspension-based (or bead-based) methods, where solid supports are used as substrates for molecular probes, are advantageous due to their 3D format. Therefore, while microarrays can offer a few million unique probes, it is conceivable that the number of probes made available by suspension-based methods can substantially increase, especially if combinatorial chemistry methods are used to synthesise a vast number of compounds. The chemical reactions occur directly on the surface of the solid supports and are not limited by diffusion while the solution phase kinetics result in faster reaction rates, therefore, assays with higher multiplex capacity can be designed. The biomolecular probes are attached to the solid supports and hybridization or other coupling reactions can take place in microfluidic channels and chips where the supports flow freely [9]. Simple modification of the assays by the straightforward selection of different molecular probes suited for specific applications is feasible with these technologies and also better quality control by batch synthesis can be achieved [10].

However, the efficient use of suspension-based technologies necessitates the development of encoding methods for the tagging of the solid supports [11]. Some of the established encoding technologies include spectrometric schemes based on the utilisation of optical beads, chromophores or fluorophores with characteristic emission spectra covalently bound to the supports [12, 13], quantum dots [14, 15] or photonic elements [16] and Raman tags [17, 18]. Graphical encoding methods where information is stored by spatial modulation of some property of a material (e.g., optically barcoded [19, 20] and patterned aluminium rods, striped metal cylindrical rods [21] and ridged or patterned particles [22]) have also been developed. Electronic encoding schemes have also been reported [23] and radio frequency memory tags offering large number of codes (more than 10^{12}) have been designed [24, 25]. However, such tags are large (several mm) and their fabrication in large numbers is expensive and slow. Physical encoding strategies that rely on the ability to discriminate between microsupports that have different physical properties (size, shape, refractive index, composition) have also been studied, but their implementation will be restricted since the number of codes will be limited [26]. A new method for the synthesis of multifunctional particles with continuous flow lithography that can provide one million codes has been reported [27]. Molecular computational identification has also been recently demonstrated as a scheme for utilizing molecular logic and computation and is able to provide millions of distinguishable codes [28].

One of the main limitations of most of the established encoding technologies is the relatively limited number of unique codes that can be provided. Fluorescent encoding for example, offers only about 100 – 1000 distinguishable tags due to spectral overlap. Utilising fluorescent semiconductor nanocrystals or quantum dots, makes feasible the generation of a few tens of thousands of distinct codes, by overcoming some of the limitations of conventional spectroscopic methods. The latter methods that use fluorophores have narrow excitation spectra which complicate the simultaneous excitation of different molecules, as well as broad emission spectra which introduce spectral overlap between different detection channels, making the extraction of quantitative information difficult. Quantum dots have size dependent tunable characteristic emission spectra that are as narrow as 20 nm in the visible region. Furthermore they can be excited using a continuous broad spectrum. These properties facilitate the simultaneous excitation of many different quantum dots each having unique emission wavelength. This results in several emission frequencies with minimal spectral overlap [29] so that about 40,000 unique codes can practically be provided.

As in the case of microarrays, optical detection techniques are the dominant encoding methods for suspension-based technologies since fluorescence-encoded microcarriers can be analysed with conventional flow cytometry instruments. However, in addition to the limitations mentioned previously, the optical detection technologies have a number of inherent disadvantages, such as the short lifetime and high cost of the fluorophores, inefficiencies due to photobleaching and limits in the implementation for parallel high throughput analysis due to cross-fluorescence and overlap between the characteristic emission spectra of the fluorophores. Moreover, most biological samples and fluids auto-fluoresce and in addition, errors can be introduced during the assay due to the interference of the encoding fluorescence with the hybridization detection fluorescence.

To take full advantage of the suspension-based technologies, it is desirable to be able to synthesise large chemical libraries of probes. With combinatorial chemistry methods, such as split and mix, chemical libraries comprising a vast and exponentially increasing number of compounds can be synthesized on solid supports with a linearly increasing number of process steps, by combining a set of a few molecular building blocks (such as the four DNA bases). The output of the synthesis is a large number of unique compounds attached to the supports, with one compound on each support; these supports however are mixed. There is, therefore, a need for the development of an encoding technology that could record the synthesis history of each compound by being able to introduce a code during the chemical reactions. Encoding the supports during the synthesis steps will result in the generation of chemical libraries of compounds with known identity. These compounds could subsequently be used as probes for high throughput biological analysis. Various methods exist for encoding during synthesis, based mainly on the addition of a chemical tag at each cycle that encodes for that particular step. These methods are, however, expensive, very labour intensive, difficult to automate and can interfere with the compound synthesis [30].

RESULTS AND DISCUSSION

Magnetic Encoding Technology

To address these limitations we are developing a novel encoding method which utilises planar digital magnetic tags that can provide a vast number of distinguishable codes, together with the ability to write the code at any stage of the assay. The magnetic encoding technique is also advantageous due to the decoupling of the encoding process from the hybridization detection process, since the former is based on magnetic detection and the latter on optical (fluorescence) detection. Magnetic detection also offers significant advantages when compared with optical detection due to the inherent ability of the technique to combine detection, as well as *actuation* of the reaction that can be carried out on the solid magnetic structures (*suspension-based technology*) thus increasing the speed of the bioassay. Magnetic elements are highly stable over time and they are not affected by chemical reagents or reasonably high temperatures, while the biological samples do not give any spurious magnetic background, since most compounds, such as DNA, cells, polymers, water, proteins are only weakly diamagnetic [31].

Magneto-resistive (MR) based sensors have been proposed as detection components for high sensitivity biosensors [32, 33, 34]. The detection of biological compounds by magnetic sensors and by the use of magnetic materials is more efficient and more sensitive than detection by optical means [35, 36]. The traditionally used labels are micron or nano- sized magnetic particles that are attached to biomolecules and the detection of the label signifies the presence of that particular molecule in the analyte. Detecting a single magnetic particle with high signal to noise ratio is very difficult for very small particles due to their low magnetic moment. Currently available magnetic sensors usually utilise mass surface coverage of beads for averaged detection of particles [38]. Several single-bead sensor geometries are being developed, such as single anisotropic magnetoresistive (AMR) rings [39], Hall crosses [40], spin valve rectangles [41] and magnetic tunnel junction (MTJ) ellipses [42]. Spin valve sensors covered with immobilized probe DNA have successfully been used to detect target DNA labelled with 250 nm magnetic beads by measuring the signal from about 100 particles positioned close to the sensor surface [43]. The use of GMR sensors integrated with microfluidic channels has also been investigated in an effort to detect magnetic particles in a continuous flow [44]. Single beads with diameters of 2.8 μm directly placed on the sensor and clusters of nanoparticles (250 nm) from droplets have been detected [45, 46]. Even though significant progress has occurred in the last few years in the field of magnetic bead detection (single particle sensing is now feasible for particles in close vicinity to the sensor) and continuous flow detection also possible for ferrofluids, the detection of single particle-labels in flow has not been achieved to date.

Controlling the position of a bead/tag relative to the sensor can be achieved by integrating the sensors into a microfluidic channel into which the beads are introduced in suspension. Beads can then be transported and positioned over the sensor by applying hydrodynamic forces or electromagnetic field gradients generated by lithographically defined current lines inside or flanking the channel [47]. Several different strategies for bead manipulation and sorting have been investigated, notably involving local magnetic field gradients generated by micro-electromagnets acting on beads or stationary suspensions [48]. We have developed an integrated microfluidic cell (IMC) comprising MR sensors, a system to focus the beads into the central flow line of the channel, a sorting gate and tapered current lines integrated with a microfluidic channel for the focusing, sorting and detection of single magnetic functionalized microbeads in solution [49].

The implementation of *magnetic encoding* techniques for high throughput bioassays (i.e. taking advantage of the ability of magnetic tags to provide distinct digital codes) has not been achieved so far, mainly due to the lack of suitable magnetic tags and the lack of sensitive sensors. In most current schemes a magnetic biosensor does not have the ability to distinguish between different populations of magnetically labeled biological compounds, since information can not be encoded onto the labels, such as the traditionally used magnetic beads. Several issues need to be addressed, such as the design, fabrication and functionalisation of the magnetic tags and their

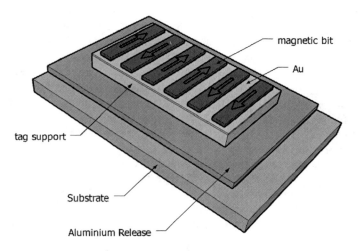

FIGURE 1. Planar multibit magnetic tag. An example of a 6 bit tag is shown here comprising a SU8 support and ferromagnetic regions in a bar code type geometry, with the arrows denoting the magnetization of each element. Each microbar is 50×20 μm^2.

detection and writing by means of sophisticated magnetic sensors and write-heads. In this paper we present the design and fabrication of planar digital tags comprising magnetic elements that are suitable for hybridization assays and we demonstrate that the magnetic elements can be encoded and decoded by means of a magnetic fluxgate sensor that is scanned over the surface of the elements which are located on a solid substrate. Traditionally, magnetic labels have been fabricated in the form of beads with diameters ranging from several nanometers to tens of microns and usually consist of a resin matrix with magnetite or other magnetic nanoparticles dispersed within it [50]. Superparamagnetic beads with a diameter of 10 microns (Bangs Lab BM547) have a magnetic moment of the order of 10^{-11} emu per bead when saturated, while ferromagnetic beads (Spherotech) have magnetic moments in the 10^{-9} emu per bead range. Ferromagnetic beads are, therefore, promising candidates for magnetic labels and in our recent work we have demonstrated the successful capture of these beads by biological cells [51].

Magnetic Microtags

We have been exploring alternative architectures to beads in the form of electrodeposited magnetic micropillars, nanowires and planar magnetic labels that can be utilized as magnetic tags for biological applications. Since the pillars are made of solid magnetic material, these structures give a much larger magnetic signal than magnetic beads [52]. A conceptually different design to the micropillars, based on planar magnetic multibit elements, has also been implemented and the utilization of these elements as magnetic tags will be described in this paper. The planar magnetic tags comprise a planar solid support and several spatially resolved switchable ferromagnetic elements (bits) deposited on its surface. A typical tag is shown schematically in Figure 1.

The ferromagnetic elements are elongated rectangles and information is stored as directions of magnetization vectors, with the spatial sequence of the bits providing the digital code. In the present specific implementation, ferromagnetic elements of 50×20 μm^2 were used. The design of the planar digital magnetic tag is generic and magnetic elements with different sizes and shapes can be used, provided that their moments are high enough to permit non contact detection. The magnetic shape anisotropy is used to achieve a preferred orientation of the magnetization of each element and their switching fields are high enough to prevent thermal perturbations from altering the encoding. The overall dimensions of the tag depend on the number of bits that is required for a particular application. Since the magnetic elements need to be decoupled so that they are written unambiguously, a minimum separation distance that is of the order of the width of the element is necessary. Additional magnetic elements may be required for error corrections and/or for controlling the orientation of the tag. The planar magnetic tags have been designed to satisfy several criteria in order to be utilized as carriers for biological assays. These criteria are *i*) the number of reactive groups on the surface to enable the attachment of the probe molecules of interest, *ii*) the ability to permit adsorption of these molecules while retaining their biological activity, *iii*) the surface area, *iv*) uniformity and monodispersion and *v*) the minimization of non-specific adsorption effects with target molecules. Some additional criteria, such as, uniform stable suspension in the absence of magnetic fields and manipulation and transport of the tags by means of external magnetic or electric fields or hydrodynamic forces or a combination thereof have also been taken into consideration when designing the tags.

The planar support has a longitudinal shape so that it can flow in microfluidic channels and can be made of a polymer material that is suitable for functionalisation, such as SU8 that has epoxy groups on its surface. Alternatively, parts of the tag can be coated with a thin gold layer that will accommodate the binding for thiol groups suitable for functionalisation with a variety of compounds. The probe compounds attached to the magnetic tags are designed to be complementary to the target compounds and the active area of the tags, defined as the total area coated with a binding molecule, can be large; resulting in high chemical sensitivity for binding assays.

To determine whether the detection of the magnetic elements is feasible using a magnetic sensor that is located at some distance away from the magnetic tags, we have fabricated magnetic elements on a wafer and measured them with a micro-fluxgate sensor. The tags were fabricated by electrodeposition, which is a fast, inexpensive and versatile fabrication method suitable for any metal that can be deposited on a conductive substrate from a solution of its ions [53]. An insulating mask on top of a conducting substrate, the template, is needed to make patterned metallic elements and was fabricated by optical lithography, while a p-doped silicon wafer coated with an evaporated Cu(35 nm)/Al(400 nm) release layer was used as the substrate. The release layer can then be dissolved away so that the tags can be collected in suspension for free flow in microchannels. The release layer is underneath the insulating resist and acts as a conductive layer which is necessary for electrodeposition to proceed. Whilst copper is a very suitable material due to its high conductivity, the addition of an

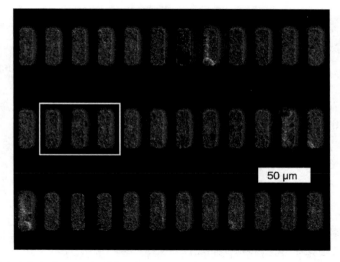

FIGURE 2. SEM micrograph of the planar magnetic tags on the substrate. The white lines schematically indicate the edges of the support of a 3 bit planar magnetic tag.

aluminium underlayer aids the release of the tags, since only a mild KOH solution is necessary to dissolve it. A quartz mask with arrays of elongated rectangles was used to pattern the samples and a PMMA resist layer was spun to a thickness of about 2 μm. The template was then cut to the desired size and a part of it at the top of the sample was removed with acetone to create the contact to the electrode. The back and the sides of the sample are protected by a chemically resistant tape and the sample exposed to the electrolytic solution (0.1 M $NiSO_4$ + 0.1 M boric acid in de-ionised water). This template was attached to an electrode and dipped into a metal solution with a counter electrode of platinum mesh and a saturated calomel (SCE) reference electrode. Soaking the sample in acetone after the deposition removes the resist and facilitates the release of the tags after functionalisation. An image of uniformly elongated tags taken after the removal of the resist with a scanning electron microscope (SEM) is presented in Figure 2.

Magnetic Micro-fluxgate Sensor

The measurements of the stray fields of the planar tags have been performed with a novel microfabricated fluxgate sensor. The fluxgate sensor is one of the most sensitive magnetic field detection devices operable at room temperature and can measure small sub-nanotesla magnetic fields (amplitude and direction) with high resolution, high sensitivity and reliable response [54]. It comprises a core of high permeability, non-linear, low loss material inside one or more coils. An alternating current applied to one of the surrounding coils periodically saturates the core in each direction. If an external magnetic field is also coupled into the core, this causes a shift in the B-H curve along the H axis axis which results in an asymmetry in the effective permeability of the core

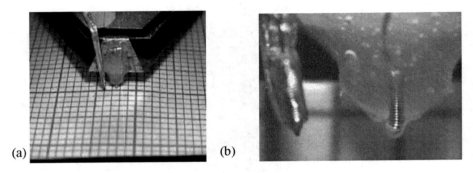

FIGURE 3. (a) Photograph of the microfabricated fluxgate sensor. (b) The write head is shown on the left of the picture and the read head on the right. Each small square on the paper shown is 1x1 mm.

between positive and negative polarities. The effect of this is to introduce even harmonics, predominantly the second harmonic, of the excitation frequency into the voltage waveform seen across the coil. A filter and demodulator circuit sensitive to these harmonics produces a signal proportional to the applied magnetic field. To improve the dynamic range and linearity of the system, the harmonic derived signal can be used in a feedback loop by passing current into the sensor coil to maintain the core at zero net flux. The feedback current then becomes the output signal.

In this work, the micro-fluxgate sensor comprises a core of the amorphous cobalt alloy $Co_{66}Si_{15}B_{14}Fe_4Ni_1$ with dimensions 1000 μm long $\times$ 35 μm wide $\times$ 5 μm thick, with the end width tapering to approximately 10 μm (Figure 3). The core is formed by RF magnetron sputtering of the as-cast amorphous ribbon and photolithography with wet chemical etching of the alloy to define the core dimensions. The sensor coil may be defined by photolithography on a glass substrate [55] or by winding with fine 35 μm diameter copper wire [56]; the latter was used in the current study. The latter method has the advantage of providing a sharp probe tip which can be easily brought into proximity with a sample surface – in this case, the sensor was encased in epoxy resin which bonded it to the end of a non-magnetic, brass cantilever. In this system a single coil provides both the excitation and sensing functions. A second coil with a 50 μm diameter iron core wound with 200 turns is mounted on an independent cantilever to enable a magnetic writing field to be produced (Figure 3b left).

The micro-fluxgate sensor is driven with an excitation current of around 50 mA at 40 MHz and the demodulation and feedback circuitry gives a measurement bandwidth up to 10 MHz. The output from the magnetic sensor system is filtered and amplified before analogue-to-digital conversion and communication back to a PC to display and store the magnetic image. The dynamic range of the system can be set from approximately 20 μT to 2 mT using a digitally programmable instrumentation amplifier and the resolution on the finest scale is about 20 nT. A PC-controlled offset of half the dynamic range can also be introduced to compensate for static stray fields

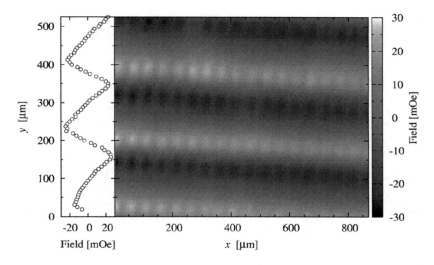

FIGURE 4. Calibrated magnetic microghaph of the elongated tags.

(the sensor is not screened from ambient fields). Under these conditions, the sensitivity of the sensor is 5000 V/T with a spectral noise density of 2 nT/√Hz at 10 kHz and an overall system bandwidth up to 2.5 MHz. The microfabricated flux gate sensor was mounted on a scanning stage and magnetic measurements of the stray field of the tags were performed with the tags placed on the stage about 50-100 μm away from the sensor head.

The samples were scanned in a raster fashion under the sensing element. The elongated magnetic elements exhibit a strong longitudinal uniaxial shape anisotropy which determines the stable equilibrium magnetic configuration of each tag. Figure 4 shows the magnetic micrograph of the sample and the stray fields from the two ends of each magnetic element (bit), which point in opposite directions, represented as a series of black and white pixels. A 2D plot of the calibrated stray field strength from an array of $50{\times}20$ μm^2 magnetic bits is presented; the detected stray magnetic field of the elements spans a range of about 60 mOe from positive to negative. The inset on this figure shows the variations of the stray magnetic field versus distance in the y direction, providing clear evidence that the magnetization state and hence the determination of each bit state, can be distinguished and read at relatively large distances from the head of the sensor. It is therefore noted that the detection of magnetic tags flowing in microfluidic channels is feasible and that the biological assay scheme proposed earlier based on a microfluidic chip with the fluxgate magnetic sensor placed in close proximity to it, but outside the channel, can be implemented without the need to integrate the magnetic sensor within the fluidic channel.

A scheme that could be used for magnetically encoded high throughput bioassays is now described. The planar tags are first fabricated on a wafer as discussed previously, while some additional photolithography steps provide the means for defining the

(a)

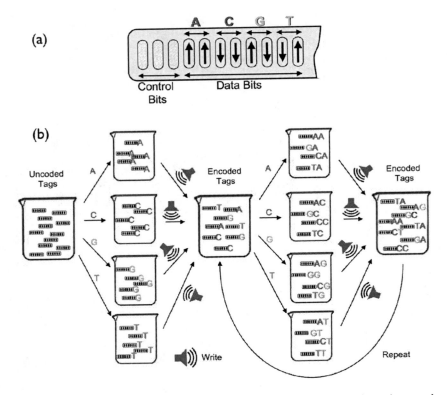

(b)

FIGURE 5. (a) A schematic diagram showing how the planar magnetic tag is used to encode oligonucleotides. (b) A conceptual diagram showing how planar magnetic tags can be used as supports for the synthesis of an encoded chemical library with the split and mix synthesis. The vessel on the left contains a vast number of blank magnetic tags in solution. These are then split into 4 vessels where the addition of a single chemical base (A, C, G or T) takes place. The tags are then carried forward to another vessel while a code that corresponds to the base that has just been synthesized is written using external magnetic flux gate sensors. The tags are then mixed again and the process is repeated as many times as necessary depending on the number of base pairs that are required (oligonucleotide length) for the probes.

substrate of the tags. The tags are then released in solution by dissolving the release layer as described above. Combinatorial chemistry methods, such as split and mix, can then be used to synthesise a *magnetically encoded* chemical library comprising very large numbers of uniquely encoded tags onto which the synthesis of the compounds is taking place. Figure 5(a) shows a schematic diagram of how the planar digital magnetic tags can encode oligonucleotides that comprise 4 chemical bases (A, C, G, or T). Each nucleotide is assigned two bits which are written sequentially onto the tag and control bits are added for data verification and error correction. It is noted that the method is easily expandable to code for more than 4 chemicals if the compound that needs to be synthesised consists of more building blocks. Figure 5(b) shows a conceptual diagram of the synthesis of an encoded chemical library of oligonucleotides with the split and mix method. External fluxgate sensors and write heads can be used to encode and decode the codes onto the tags as they flow past in fluidic channels.

CONCLUSIONS

The design and fabrication of magnetic tags that can be used for labelling biological compounds have been discussed in this paper. The magnetization direction of the magnetic elements of each tag provides the mechanism for encoding unique codes that correspond to the identity of the biological compound with which the tag is functionalized. A plethora of unique codes can be provided by using several magnetic elements (bits) on every tag. We have demonstrated that magnetic elements (bits) with dimensions of 50×20 μm^2 can be decoded by an external microfabricated fluxgate magnetic sensor that is placed in relatively close proximity (50-100 μm away) to the surface of the tags. The tags can be suspended in a viscous solution so that they can then be transported over the sensor inside microfluidic channels. A magnetic lab-on-a-chip device can then be designed that will be used for multiplex biological analysis. If combinatorial chemistry methods are implemented, vast numbers of molecular probes can be synthesise directly on the surface of the tags. We believe that by utilising the ability of the magnetic tags to be encoded in flow while the synthesis of the chemical library is taking place, as well as their capability to adopt a plurality of unique magnetic configurations and hence provide a plurality of unique codes, the design of a magnetic device, that will enable magnetically encoded suspension-based ultra high throughput biochemical analysis, is feasible.

ACKNOWLEDGMENTS

This paper is dedicated to the memory of Professor J.A.C. Bland whose recent death was a great shock and who is greatly missed. Professor Bland initiated and managed this interdisciplinary research project ("4 Billion Bases a Day"), funded by the EPSRC Basic Technology Programme. The EPSRC is acknowledged for providing financial support.

REFERENCES

1. V.G. Cheung, M. Morley, F. Aguilar, A. Massimi, R. Kucherlapati and G. Childs, *Nature Genetics* **21**, 15 (1999).
2. M. Schena, D. Shalon, R. Davis and P. Brown, *Science* **270**, 467 (1995).
3. T.R. Hughes, M. Mao, A.R. Jones, J. Burchard, M.J. Marton, K.W. Shannon, S.M. Lefkowitz, M. Ziman, J.M. Schelter, M.R. Meyer, S. Kobayashi, C. Davis, H.Y. Dai, Y.D.D. He, S.B. Stephaniants, G. Cavet, W.L. Walker, A. West, E. Coffey, D.D. Shoemaker, R. Stoughton, A.P. Blanchard, S.H. Friend and P.S. Linsley, *Nature Biotechn.* **19**, 342 (2001).
4. D.J. Lockhart and E.A. Winzeler, *Nature* **405**, 827 (2000).
5. M. Jayapal and A.J. Melendez, *Clinical and Exper. Pharmacology and Physiology* **33**, 496 (2006).
6. D.B. Allison, X. Cui, G.P. Page and M. Sabripour, *Nature Reviews* **7**, 55 (2006).
7. F. Bier and F. Kleinjung, *Fresenius J. Anal. Chem.* **371**, 151 (2001).
8. M.R. Henry, P. Wilkins Stevens, J. Sun and D.M. Kelso, *Anal. Biochem.* **276**, 204 (1999).
9. B.J. Battersby, G.A. Lawrie and M. Trau, *Drug Discov. Today* **6** (12), S19 (2001).

10. J.P. Nollan and L.A. Sklar, *Trends Biotechnol.* **20**, 9 (2002).
11. K. Braeckmans, S.C.D. Smedt, C. Roelant, M. Leblans, R. Pauwels and J. Demeester, *Nature Reviews: Drug Discovery* **1**, 447 (2002).
12. J. Raez, D.R. Blais, Y. Zhang, R.A. Alvarez-Puebla, J.P. Bravo-Vasquez, J.P. Pezacki and H. Fenniri, *Langmuir* **23** (12), 6482 (2007).
13. B.J. Battersby, D. Bryant, W. Meutermans, D. Matthwews, M.L. Smythe and M. Trau, *J. Am. Chem. Soc.* **122**, 2138 (2000).
14. M. Han, X. Gao, J.Z. Su and S. Nie, *Nature Biotechnology* **19**, 631 (2001).
15. P.S. Eastman, W. Ruan, M. Doctorelo, R. Nuttall, G. de Feo, J.S. Park, J.S.F. Chu, P. Cooke, J.W. Gray, S. Li and F.F. Chen, *Nano Lett.* **6** (5), 1059 (2006).
16. F. Cunin, T.A. Schmedake, J.R. Link, Y.Y. Li, J. Koh, S.N. Bhatia and M.J. Sailor, *Nature Mater.* **1**, 39 (2002).
17. X. Su, J. Zhang, L. Sun, T.W. Koo, S. Chan, N. Sundararajan, M. Yamakawa and A.A. Berlin, *Nano Lett.* **5**, 49 (2005).
18. Y.W.C. Cao, R.C. Jin and C.A. Mirkin, *Science* **297**, 1536 (2002).
19. G. Galitonov, S. Birtwell, N. Zheludev and H. Morgan, *Optics Express* **14**, 1382 (2006).
20. S. Banu, S. Birtwell, G. Galitonov, Y. Chen, N. Zheludev and H. Morgan, *J. Micromech. Microeng.* **17**, S116 (2007).
21. S.R. Nicewarner-Pena, R. Griffith Freeman, B.D. Reiss, L. He, D.J. Pena, I.D. Walton, R. Cromer, C.D. Keating and M.J. Natan., *Science* **294**, 137 (2001).
22. Z.L. Zhi, Y. Morita, Q. Hasan and E. Tamiya, *Anal. Chem.* **75**, 4125 (2003).
23. D.K. Wood, G.B. Braun, J.L. Fraikin, L.J. Swenson, N.O. Reich and A.N. Cleland, *Lab on a Chip* **7** (4), 469 (2007).
24. R.F. Service, *Science* **270**, 577 (1995).
25. M.M. Miller, G.A. Prinz, S-F. Cheng and S. Bounnak, *Appl. Phys. Lett.* **81**, 2211 (2002).
26. H. Fenniri, L.H. Ding, A.E. Ribbe and Y. Zyrianov, *J. Am. Chem. Soc.* **123**, 8151 (2001).
27. D.C. Pregibon, M. Toner and P.S. Doyle, *Science* **315**, 1393 (2007).
28. A.P. de Silva, M.R. James, B.O.F. McKinney, D.A. Pears and S.M. Weir, *Nature Mater.* **5** (10), 787 (2006).
29. M. Bruchez, M. Moronne, P. Gin, S. Weiss and A.P. Alivisatos, *Science* **281**, 2013 (1998).
30. A.W. Czarnik, *Proc. Natl. Acad. Sci.* **94**, 12738 (1997).
31. V. Joshi, G. Li, S.X. Wang and S. Sun, *IEEE Trans. Magn.* **40** (4), 3012 (2004).
32. D.R. Baselt, G.U. Lee, M. Natesan, S.W. Metzger, P.E. Sheehan and R.J. Colton, *Biosensors and Bioelectronics* **13**, 731 (1998).
33. G.X. Li, V. Joshi, R.L. White, S.X. Wang, J.T. Kemp, C. Webb, R.W. Davis and S.H. Sun, *Appl. Phys.* **93**, 7557 (2003).
34. R.L. Edelstein, C.R. Tamanaha, P.E. Sheehan, M.M. Miller, D.R. Baselt, L.J. Whitman and R.J. Colton, *Biosens. Bioelectron.* **14**, 805 (2000).
35. H.J. Kim, S.H. Jang, K.H. Oh, T.S. Kim and K.Y. Kim, *Phys. Status Solidi A-Appl. Res.* **201**, 1961 (2004).
36. S.J. Park, T.A. Taton and C.A. Mirkin, *Science* **295**, 1503 (2002).
37. M. Tondra, M. Porter and R.J. Lipert, *J. Vac. Science and Technology* A **18** (4), 1125 (2000).
38. M. Megens and M. Prins, *J. Magn. Magn. Matt.* **293**, 702 (2005).
39. K. Miller, *The Scientist* **16**, 52 (2002).
40. L. Ejsing, M.F. Hansen, A.K. Menon, H.A. Ferreira, D.L. Graham and P.P. Freitas, *J. Magn. Magn. Mat.* **293**, 677 (2005).
41. H.A. Ferreira, D.L. Graham, P.P. Freitas and J.M.S. Cabral, *J. Appl. Phys.* **93** (10), 7281 (2003).
42. W. Shen, X. Liu, D. Mazumdar and G. Xiao, *Appl. Phys. Lett.* **86**, 253901 (2005).
43. D.L. Graham, H.A. Ferreira, N. Feliciano, P.P. Freitas, L.A. Clarke and M.D. Amaral, *Sens. Actuators B* **107**, 936 (2005).
44. N. Pekas, M.D. Porter, M. Tondra, A. Popple and A. Jander, *Appl. Phys. Lett.* **85**, 4783 (2004).
45. L. Ejsing, M.F. Hansen, A.K. Menon, H.A. Ferreira, D.L. Graham and P.P. Freitas, *Appl. Phys. Lett.* **84**, 4729 (2004).
46. P.A. Besse, G. Boero, M. Demierre, V. Pott and R. Popovic, *Appl. Phys. Lett.* **80**, 4199 (2002).
47. Z. Jiang, J. Llandro, T. Mitrelias and J.A.C. Bland, *J. Appl. Phys.* **99** (8), 08S105 (2006).

48. L. Lagae, R. Wirix-Speetjens, J. Das, D. Graham, H. Ferreira, P.P. Freitas, G. Borghs and J. de Boeck, *J. Appl. Phys.* **91** (10), 7445 (2002).
49. T. Mitrelias, Z. Jiang, J. Llandro and J.A.C. Bland, "Magnetic Devices for Biological Analysis", *Techn. Proc. of the 2006 Nanotechnology Conf., Boston, USA,* **Vol. 2**, pp. 256 (2006).
50. Q.A. Pankhurst, J. Connolly, S.K. Jones and J. Dobson, *J. Phys. D.* **36**, R167 (2003).
51. T. Mitrelias, J. Palfreyman, Z. Jiang, J. Llandro, J.A.C. Bland, R.M. Sanchez-Martin and M. Bradley, *J. Mag. Magn. Mat.* **310** (2), 2862 (2007).
52. J.J. Palfreyman, F. van Belle, W.S. Lew, T. Mitrelias, J.A.C. Bland, M. Lopalco and M. Bradley, *IEEE Trans. Magnet.* **43** (6), 2439 (2007).
53. W. Schindler, O. Schneider and J. Kirschner, *J. Appl. Phys.* **81**, 3915 (1997).
54. P. Ripka, *Sens. and Actuat. A* **33**, 129 (1992).
55. P. Robertson, *Electron. Lett.* **36** (4), 331 (2000).
56. P. Robertson, *Electron. Lett.* **33** (5), 396 (1997).

High Throughput Biological Analysis Using Multi-bit Magnetic Digital Planar Tags

B. Hong*, J.-R. Jeong†, J. Llandro*, T.J. Hayward*, A. Ionescu*, T. Trypiniotis*, T. Mitrelias*, K.P. Kopper*, S.J. Steinmuller* and J.A.C. Bland*

*Cavendish Laboratory, University of Cambridge, J J Thomson Avenue, Cambridge, CB3 0HE, United Kingdom
†Department of Materials Engineering, Chungnam National University, Daejeon 305-764, South Korea

Abstract. We report a new magnetic labelling technology for high-throughput biomolecular identification and DNA sequencing. Planar multi-bit magnetic tags have been designed and fabricated, which comprise a magnetic barcode formed by an ensemble of micron-sized thin film $Ni_{80}Fe_{20}$ bars encapsulated in SU8. We show that by using a globally applied magnetic field and magneto-optical Kerr microscopy the magnetic elements in the multi-bit magnetic tags can be addressed individually and encoded/decoded remotely. The critical steps needed to show the feasibility of this technology are demonstrated, including fabrication, flow transport, remote writing and reading, and successful functionalization of the tags as verified by fluorescence detection. This approach is ideal for encoding information on tags in microfluidic flow or suspension, for such applications as labelling of chemical precursors during drug synthesis and combinatorial library-based high-throughput multiplexed bioassays.

Keywords: Magnetic tagging, High-throughput bioassay, Remote encoding.
PACS: 85.70.Ay, 85.70.Kh, 87.85.Rs, 87.85.Qr

INTRODUCTION

The development and application of microarrays has become an important trend in clinical diagnosis, DNA sequencing, drug screening and discovery, and other types of biological research [1]. Microarrays identify hybridization of a particular fluorescently labelled biochemical probe (usually a short DNA sequence or recognition protein) to the unknown biomolecule (the target) by observation of a spatially resolved fluorescence signal. However, it is difficult to both automate and extend the detection capability and analysis of the results must be done not in real time, but rather by postprocessing significant amounts of stored raw data. Therefore, very recent efforts have concentrated on performing assays in microfluidic flow cells using microfabricated labels which identify by a characteristic optical signature the single species of biomolecular probe to which they are specifically bound. This method allows flexible, automated, high-throughput analysis, whose detection capabilities can be extended simply by increasing the size of the library of probes rather than the number of sites in the microarray. Optical encoding methods reported include spectrometric [2–4] (fluorophores, quantum dots and Raman tags), and image-based or graphical [5, 6] identification. Fluorescent labelling in particular has become the current dominant technology in the field of molecular

CP1025, *Biomagnetism and Magnetic Biosystems Based on Molecular Recognition Processes*
edited by J. A. C. Bland and A. Ionescu

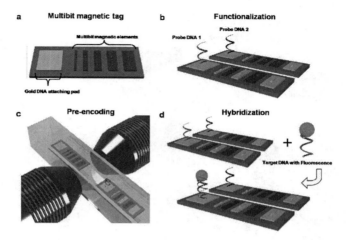

FIGURE 1. Schematics diagrams for encoding and hybridization of multi-bit digital magnetic tags. (a.) Multibit magnetic tag. (b.) Magnetic tags are functionalized by attaching the probe DNA to the corresponding tag. (c.) Magnetic elements in sequence of aspect ratios are individually addressable and they are pre-encoded depending on the probe DNAs. (d.) Only the tags that show a positive match between the probe and the complementary target DNA will show fluorescence.

identification. However, spectral overlap limits multiplexing and errors are introduced by interference between signatures used to verify probe functionalisation of the labels, hybridisation between the probes and the targets, and finally to identify the probe (and thus the target) [6]. The ideal combination of sensor and label would independently quantify the ability of probe molecules to bind both to the labels and targets, and also identify the targets from the hybridisation results in a fully automated way. Magnetically encoded and detected labels [7] or "magnetic tags" are a natural choice, due to their stability with respect to time, temperature, and reagent chemistry [8] and the lack of magnetic background generally found in biological samples. Furthermore, decades of research on hard disk technology leads naturally to the concept of encoding more than one bit of information on a magnetic tag, leading to a multiplexing power unavailable to optical methods.

DESIGN OF HIGH THROUGHPUT BIOLOGICAL ANALYSIS SYSTEM

Here, we present a new remote encoding and decoding scheme for micrometer-sized multi-bit magnetic tags which utilizes magnetic digital planar tags as carriers for bio-molecules, and micron-sized magnetic barcodes to encode a unique signature on each label for performing multiplexed high-throughput bioassays. We also describe in later sections, the results of digital planar tag fabrication and releasing, as well as a demonstration of remote encoding, reading and decoding of a 5-bit digital planar tag using a

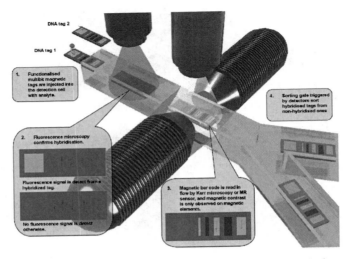

FIGURE 2. Overview of high-throughput multiplexed biological analysis system.

full field Kerr Microscope.

The general design of a digital planar tag is shown in Fig. 1. a. The design contains three sections: Multi-bit magnetic elements with individually different coercivity for remote signature encoding, a gold attaching pad for immobilizing the DNA probes and a backbone made of fully cross linked SU8 polymer for high chemical resistance. The fabrication of digital planar tags is described in the following section. Digital planar tags are functionalized with different DNA probes in suspension and followed by remote encoding by external fields either in suspension or a micro-fluidic channel, as described in Fig. 1. b and c. The analyte containing both DNA probes is hybridized with the fluorescently labeled target DNA and only the tags that show a positive match between the probe and the complementary target DNA will show fluorescence, as shown in Fig. 1. d.

Magnetic barcodes are specially engineered to exhibit individually different coercivities to an applied external field. This was achieved by controlling the aspect ratio of the magnetic elements which dominates the shape anisotropy of micro meter sized particles. Magnetic elements with two stable states along its long axis can, therefore, be addressed individually by a varying external field, as shown in Fig. 6, the details will be described in the following section.

In Fig. 2, we demonstrate the detection method of digital planar tags in a microfluidic cell. Fluorescence microscopy was utilized to confirm the hybridizion, full field Kerr microscopy to decode the signature and a copper strapline to induce a field gradient to sort the digital planar tags [9]. The digital planar tags can also be sorted into other microfluidic devices for other processes after detection. In this way, the highly parallel screening of biological samples can be performed very rapidly, since the decoding of the hybridized magnetic tags takes place in a few tenth of microseconds. The factors that introduce errors or limit throughput of such screening techniques are associated only with the biochemical processes, such as the time necessary for hybridization, but are not associated with the magnetic encoding or decoding processes.

In the writing process, a negative external magnetic field H_1 ($H_1 \ll$ -h_{1c}) strong enough to align all the elements in the series is first applied to initialize the tag, where h_{ic} with i=1,...,5 is defined as the respective coercive field necessary to reverse the magnetic state of ith element. Then the ensemble of the tags can be defined by the vector $M(H_1)$=(1,1,1,1,1) (Fig. 3. I). If a positive field H_2 (h_{2c}<H_2<h_{1c}) is next applied, sufficient to switch all elements except the last element with the highest switching field, then the ensemble becomes $M(H_2)$=(1,0,0,0,0) as shown in Fig. 3. I. If a field h_{3c}<H_3<h_{2c} is applied, it switches the third element without disturbing the states of the previously written bits and the ensemble becomes $M(H_3)$=(1,0,1,1,1). The process can then be repeated for the next softest moment using a field h_{4c}<H_4<h_{3c} and so on, until all elements in the ensemble have been reversed into a desired state associated to a certain probe, e.g. $M(H_5)$=(1,0,1,0,1). Therefore, it is possible to encode any state by applying a series of decreasing field pulses of arbitrary signs, in which for each bit the magnitude of the pulse selects the bit and its sign determines the magnetization direction of each bit, as shown in Fig. 3. I. Decreasing strength ensures that the hardest layer is written only once; where the softest layer has to be reversed N-times before the final state is achieved. It is worthwhile to note that the advantage of this approach is that the fields applied are not local but global which does not require close proximity and careful alignment. This is ideal for writing codes on tags in microfluidic flow or a large quantity of the same code on tags in suspension.

FABRICATION OF SU8 ENCAPSULATED DIGITAL PLANAR TAGS

The magnetic planar tags consisting of magnetic elements in between two SU8 backbones were fabricated and released. This fabrication procedure is shown in Fig. 4. A silicon substrate was first coated with a 400 nm-thick aluminium sacrificial layer by thermal evaporation (Fig. 4. a) which is dissolvable in tetramethylammonium hydroxide (a.k.a MF319). A layer of 1 μm-thick SU8 2002 negative photoresist was spun onto the substrate and patterned by UV photolithography (Fig. 4. b), the unexposed areas were removed by developing in methyl isobutylketone (MIBK) for 120 sec. A permanent structure was obtained by hardbaking the sample at 150°C for one hour to fully crosslink the polymer.

Magnetic barcode elements were afterwards defined by DUV photolithography in a layer of 1 μm-thick polymethyl methacrylate (PMMA) photoresist which was spun onto the substrate consisting of SU8 backbones and an aluminium sacrificial layer (Fig. 4. c). The exposed sample was developed for 120 sec. in a solution of 1:3 (v/v) MIBK: Isopropanol (IPA) (Fig. 4. d and e). Metallic layers consisting of 10 nm Cr /20 nm Permalloy(Py) /10 nm Cr were then grown by molecular beam epitaxy (MBE) in an ultra-high vacuum chamber with a base pressure of 1×10^{-10} mBar. Cr was used to promote the adhesion between the magnetic thin film and the SU8 backbone (Fig. 4. f). Another layer of 1 μm-thick SU8 polymer was patterned through the same procedure to fully encapsulate and protect the magnetic elements (Fig. 4. g and h). Finally, the

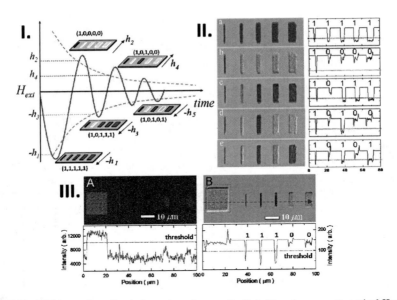

FIGURE 3. Writing and reading information of magnetic digital tags by magneto-optical Kerr effect. (I.) Schematics of writing the barcode by applying an external AC magnetic field. (II.) Reading / decoding by magneto-optical Kerr effect microscopy. (III.) Confirmation of the binding on the Au pad by fluorescent microscopy.

patterning process was repeated for the 10 nm Cr/ 100 nm Au layer to immobilize the DNA probe. Digital planar tags were then released into tetramethylammonium hydroxide solution by etching the aluminium sacrificial layer (Fig. 4. i), followed by water dilution cycles using a centrifuge to reach a tag concentration of approximately 1M

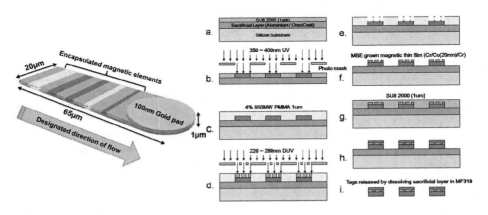

FIGURE 4. Schematics of a SU8 encapsulated digital planar tag and the fabrication procedures.

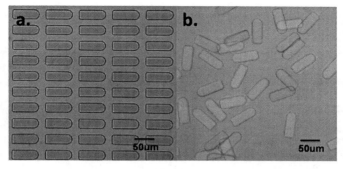

FIGURE 5. Optical microscopic image of SU8 tags. (a.) SU8 tags before releasing. (b.) Released SU8 tags in water.

/ 20ml and a solvent concentration of <0.01%. Fig. 5 shows the image of SU8 backbones before (a) and after release (b) into water.

EXPERIMENTAL RESULTS

In this study, the control of the switching fields is engineered by using the well known effects of shape anisotropy [10–12]. Arrays of micron-sized magnetic elements of 20 μm height with different aspect ratios (aspect ratio = height / width) and shapes were fabricated and studied to optimize the magnetic properties of digital planar tags. The switching distributions of these magnetic elements were measured statistically using a magneto-optical Kerr microscope, which is capable of imaging the magnetic domains with high resolution. The width of the measured elements were 1 μm, 2 μm, 3 μm, 4 μm, 6 μm and 8 μm, respectively, which gave a wide range of aspect ration between 1:20 to 1:2.5. The coercive fields of more than 60 elements were measured for each type and the data was analyzed statistically as a Gaussian distribution. Among the rectangular, hexagonal and elliptical elements that have been studied, rectangular elements have shown the most desirable switching behavior with excellent reproducibility and almost linear relationship between coercive field and aspect ratio over the measurement range where the aspect ratio is greater than five, as shown in the Fig. 6. The nonlinearity was observed for larger elements with aspect ratio less than five. This was due to the formation of double vortex states and their domination in magnetic reversal process rather than the effect of shape anisotropy. The other shapes have shown a non-linear relationship between coercive field and aspect ratio as well as a higher variation between individual elements due to the lithographical limitations as well as the double vortex states, although a higher overall switching field was observed.

For the remote sensing method, we utilized a full-field magneto-optical Kerr microscope to observe the magnetic domain configuration [13, 14], due to its high sensitivity to surface magnetization direction. The Kerr microscope consists of an optical polarizing microscope, capable of imaging magnetic domains by the in-plane longitudinal magneto-optical Kerr effect. A CCD camera captures the domain image in the form of

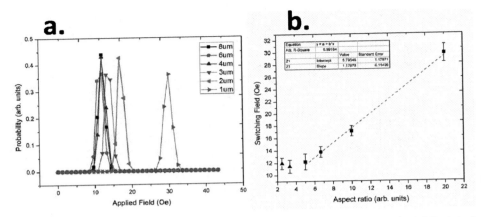

FIGURE 6. Switching distribution of rectangular elements with different aspect ratio. (a.) Statistical measurement of the switching distribution of rectangular elements with widths between 1 to 8 μm. (b.) Switching field versus aspect ratio for rectangular elements. The dashed line represents a linear fit to the data with aspect ratio greater than 5.

the array of the Kerr intensity as measured from the pixels. By storing the domain images and tracing the Kerr intensity variation for every individual pixel, it is possible to obtain an array of the local magnetic information.

In Fig. 3. II, we demonstrated the encoding of a 5-bits magnetic tag as corresponding to the writing procedure as described in an earlier section and indicated in Fig. 3. I. The magnetic barcode was encoded by a varying externally applied magnetic field as shown in Fig. 3. I. This was than imaged by Kerr microscopy to confirm the signature, as shown in Fig. 3. II. a to e. In Fig. 3. III, we demonstrate the detection of the binding and decoding of magnetic tags by using the fluorescent and Kerr microcopies after the hybridization of the oligonuclotide sequences. Figures 3. III. A and 3. III. B show the spatially resolved fluorescence and magnetic Kerr signals obtained by the confocal and Kerr microscopies with corresponding line intensity profiles. The line profiles of the Cy3-fluorescent and magneto-optical images clearly show the hybridization of the target analyte with the probe and defines the states of the magnetic tag, $M=(1,1,1,0,0)$, which contains the information of the probe materials. These results demonstrate the feasibility of remote encoding and decoding of multi-bit magnetic tags and the detection of the target bound of analyte. The power of this approach is that both encoding and decoding methods are performed globally which is suitable for the high-throughput multiplexed bioassay.

CONCLUSION

In conclusion, we have demonstrated the engineering and fabrication of planar multi-bit magnetic tags. This technique has the advantages over other high throughput biological assays in two main aspects. The numbers of unique codes that can be generated is potentially very large with a vast number of magnetic bits and the ability to encode information on the tag. This means that the tags can be used as microlabels for the synthesis of very large and diverse encoded chemical libraries (synthesized for instance by combinatorial split and mix methods), where the synthesis history of each compound library member can be recorded. The number of unique codes can be extended by increasing the number of magnetic elements fabricated onto the polymer backbone. In addition, the magnetic tags can be functionalized with a variety of biological probes, such as antibodies, proteins, or even entire biological cells [15], depending on the nature of the biological assay that needs to be performed.

ACKNOWLEDGMENTS

This work was supported by the EPSRC 4G Basic Technology Project. The author (B. Hong) would like to thank Professor J.A.C Bland for the financial support.

REFERENCES

1. S.P. Fodor, R.P. Rava, X.C. Huang, A.C. Pease, C.P. Holmes and C.L. Adams, *Nature* **364**, 555 (1993).
2. M. Han, X. Gao, J.Z. Su and S. Nie, *Nat. Biotechnology* **19**, 631-635 (2001).
3. P.S. Eastman, W. Ruan, M. Doctolero, R. Nuttall, G.D. Feo, J.S. Park, J.S.F. Chu, P. Cooke, J.W. Gray, S. Li and F.F Chen, *Nano Letters* **6**, 1059-1064 (2006).
4. Y.W.C. Cao, R.C. Jin and C.A. Mirkin, *Science* **297**, 1536-1540 (2002).
5. S.R. Nicewarner-Peña, R.G. Freeman, B.D. Reiss, L. He, D.J. Peña, I.D. Walton, R. Cromer, C.D. Keating and M.J. Natan, *Science* **294**, 137-141 (2001).
6. D.C. Pregibon, M. Toner and P.S. Doyle, *Science* **315**, 1393-1396 (2007).
7. D.L. Graham, H.A. Ferreira and P.P. Freitas, *Trends in Biotechnology* **22** (9), 455-462 (2004).
8. V. Joshi, G. Li, S.X. Wang and S. Sun, *IEEE Trans. Magn.* **40**, 3012-3014 (2004).
9. Z. Jiang, J. Llandro, T. Mitrelias and J.A.C. Bland, *J. Appl. Phys.* **99**, 08S105 (2006).
10. C.C. Change, W.S Chung, Y.C. Change and J.C. Wu, *IEEE Trans. Magn.* **41**, 947 (2005).
11. E. Girgis, J. Schelten, J. Janesky. S. Tehrani and H. Goronkin, *Appl. Phys. Lett.* **76**, 3780 (2000).
12. P. Vavassori, O. Donzelli, V. Metlushko, M. Grimsditch, B. Ilic, P. Neuzil and R. Kumar, *J. Appl. Phys.* **88**, 999 (2000).
13. A. Hubert and R. Schaefer,"Domain observation techniques" in *MagneticDomains*, edited by A. Hubert and R. Schaefer, Berlin: Springer, 1998, pp. 11-104.
14. S.-B. Choe, D.-H. Kim, Y.-C. Cho, H.-J. Jang, K.-S. Ryu, H.-S. Lee and S.-C. Shin, *Rev. Sci. Instrum.* **73**, 2910 (2002).
15. T. Mitrelias, Z. Jiang, J. Llandro and J.A.C. Bland, *J. Mag. Magn. Mat.* **310** (2), 2862 (2007).

Magnetically Controlled Shape Memory Behaviour – Materials and Applications

A.P. Gandy[a], A. Sheikh[a], K. Neumann[a], K.-U. Neumann[a], D. Pooley[a] and K.R.A. Ziebeck[a]

[a]Department of Physics, Loughborough University, LE11 3 TU, UK.

Abstract. For most metals a microscopic change in shape occurs above the elastic limit by the irreversible creation and movement of dislocations. However a large number of metallic systems undergo structural, martensitic, phase transformations which are diffusionless, displacive first order transitions from a high-temperature phase to one of lower symmetry below a certain temperature T_M. These transitions which have been studied for more than a century are of vital importance because of their key role in producing shape memory phenomena enabling the system to reverse large deformations in the martensitic phase by heating into the austenite phase. In addition to a change in shape (displacement) the effect can also produce a force or a combination of both. Materials having this unique property are increasing being used in medical applications – scoliosis correction, arterial clips, stents, orthodontic wire, orthopaedic implants etc. The structural phase transition essential for shape memory behaviour is usually activated by a change in temperature or applied stress. However for many applications such as for actuators the transformation is not sufficiently rapid. Poor energy conversion also limits the applicability of many shape memory alloys. In medicine a change of temperature or pressure is often inappropriate and new ferromagnetic materials are being considered in which the phenomena can be controlled by an applied magnetic field at constant temperature. In order to achieve this, it is important to optimise three fundamental parameters. These are the saturation magnetisation σ_s, the Curie temperature T_c and the martensitic temperature T_M. Here, σ_s is important because the magnetic pressure driving the twin boundary motion is $2\sigma_s H$. Furthermore the material must be in the martensitic state at the operating temperature which should be at or above room temperature. This may be achieved by alloying or controlling the stoichiometry. Recently new intermetallic compounds based on the ferromagnetic prototype Ni_2MnGa have been discovered which offer the possibility of controlling the structural phase transition by a magnetic field, hence opening up new possible applications particularly in the field of medicine. The properties of these new materials will be presented and their suitability for applications discussed.

Keywords: Magnetic shape memory, Magnetism, Neutron scattering, Phase transition.
PACS: 61.66.Dk; 81.30.Kf; 75.90.+w

INTRODUCTION

Clinical procedures have in recent years made rapid progress. This is in part due to the development of new materials for use as implants and instruments permitting non-invasive surgery [1, 2]. Owing to their inert properties, precious metals were initially chosen for medical and dental applications. However a wide range of materials e.g. metals, alloys, ceramics, polymers, carbon composites as well as natural materials

CP1025, *Biomagnetism and Magnetic Biosystems Based on Molecular Recognition Processes*
edited by J. A. C. Bland and A. Ionescu
© 2008 American Institute of Physics 978-0-7354-0547-9/08/$23.00

such as bone are now being employed. This article will focus on metallic alloy systems. For medical and dental applications it is not sufficient to establish the mechanical properties and integrity of such systems but also their biocompatibility. When *in situ* for example, steels corrode, nickel is toxic and chromium cobalt alloys wear. Some of these deficiencies can be overcome or minimised by combining several materials e.g. metals and polymers for hip prostheses or TiO_2 coating to reduce corrosion. A further problem is the fixation of implants which can involve mechanical pinning or the use of adhesives. Consequently several different materials can be involved giving rise to complex compatibility, possible interface problems and long term reliability. These problems and those of rejection and trauma can be minimised if the device or implants are constructed from a single smart material that can respond to their environment. Shape memory alloys are one such class. In particular NiTi (~55% wt Ni, commercial name nitinol) is providing a number of solutions: micromanipulators that mimic muscles, fixation of artificial joints, cardiovascular devices such as stents and sieves, orthodontic drills and braces, surgical instruments such as endoscopes and ophthalmic spectacle lenses.

SHAPE MEMORY BEHAVIOUR

Materials exhibiting shape memory effect [3, 4] can be formed at one temperature T_F, then cooled to a low-temperature T_D and plastically deformed; on re-heating to T_F they will regain their original shape. In addition to a change in shape (displacement) the effect can also produce a force or a combination of both. The origin of this effect is a structural phase transition which must occur at a temperature T_M intermediate between T_F and T_D and is therefore usually activated by a change in temperature or applied stress, Figure 1. Although it is usual for the shape to be formed in the high-temperature parent phase it is possible to have a two-way shape memory effect in which the shape in the martensitic phase is remembered. Here only the more widely employed one-way mechanism will be discussed. If the materials are to be of practical use it is important that the hysteresis associated with the first order structural phase transition is as small as possible. The transformation process is characterised by four temperatures, M_s and M_f, the martensitic start and finish temperatures on cooling and similarly for the heating cycle A_s and A_f which are the austenite start and finishing temperatures. For materials that undergo a thermo-elastic transition the hysteresis is usually defined as $\Delta T = M_s - A_f$ is of the order of 10 K. For convenience the transition temperature will taken as the mid point and referred to as T_M. On cooling below T_M crystallographic domains, known as variants, are usually formed as a result of the lower crystal symmetry, Figure 1. The application of external stress can cause the variants to rearrange through twin motion ultimately creating a specimen with a single variant. The stress may also influence the basic structure of the martensite which is often modulated as shown later in the article in Figure 7.

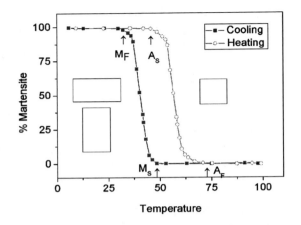

FIGURE 1. A schematic representation of the austenite (A) - martensite (M) phase transition showing the hysteresis arising from the heating and cooling cycles. The structure of the high symmetry A phase is represented by the square and the low-temperature matensitic structure by the rectangles, indicating the presence of variants.

MECHANICAL PROPERTIES OF SHAPE MEMORY ALLOYS

Super-Elasticity

A schematic phase diagram in the stress-temperature plane is shown in Figure 2. If the stress is increased at constant temperature above the phase transition A_f, then the resulting stress–strain dependence is shown as the upper curve in Figure 3. Initially there is a linear stress-strain dependence until the phase boundary is reached.

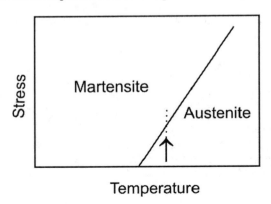

FIGURE 2. The martensitic phase transformation in the stress-temperature plane showing the critical stress line.

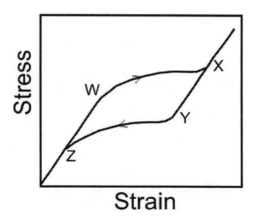

FIGURE 3. The stress-strain dependence above A_f along the path indicated by the arrow in Figure 2.

Increasing the stress further causes a rearrangement of the variants until a single variant state is obtained and a linear response occurs again. The unloading curve gives rise to the reverse procedure and hysteresis. The non linear stress-strain dependence gives rise to strains larger than those normally observed above the elastic limit for which an essentially constant stress occurs for a wide range of strain. In this region the force is not dependent on strain but only on temperature. The general shape of the stress-strain variation shown in Figure 3 is very similar to many natural bio-materials e.g. bone, tendon etc hence aiding compatibility. Since body temperature is essentially constant this mechanical characteristic has important applications. Dental braces used for aligning teeth, is one such application. As the teeth move the strain changes but the force remains constant thus avoiding the need for the braces to be readjusted during treatment. More recently super-elasticity has been used to construct 'unbreakable' spectacle frames. These frames can be twisted or bent and so momentarily deformed but on release they return to their original shape.

Shape Memory

There are two types of shape memory behaviour, here only the one-way effect will be considered. The desired shape is formed in the austenite phase. If on cooling into the martensitic phase the material is plastically deformed it remains in this state unless reheated back into the austenite phase where it regains its original shape. The stress-strain variation of this cycle in the martensitic phase is shown schematically in Figure 4. As the stress initially increases there is an approximately linear dependence on strain until the variants start to rearrange and eventually a single variant state is achieved. On unloading, the system follows the path Y-Z and remains in a deformed state once the stress has been completely removed. The original shape is only obtained by heating above A_f. This process is referred to as free recovery and is employed in stents and vena cava filters. For example the shape of the filter is formed in the austenite phase it is then cooled into the martensitic phase and drawn into a cannular which is used to position the filter in the vascular system. Once in position the cooling

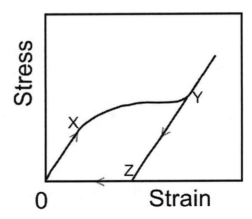

FIGURE 4. The stress-strain dependence of a shape memory alloy in the martensitic phase.

is stopped and the filter extruded. The increase in temperature produced by the body causes the filter to heat back into the austenite phase and to regain its shape.

If the deformed structure in the martensitic phase is constrained, then on heating a temperature dependent recovery force is produced. This is known as constrained recovery, which can be used for clamping and fixations e.g. dental root implants, prostheses, staples and bone anchors etc.

For some applications particularly those of a clinical nature the initiation of the shape memory effect by a change in temperature or stress may be inappropriate. Therefore there has been considerable research in recent year to find systems in which the process can be controlled by a magnetic field at constant temperature and stress.

FERROMAGNETIC SHAPE MEMORY ALLOYS

Ferromagnetic shape memory alloys show large magnetic field induced strains of over 5% by the rearrangement of twin variants in the martensitic phase [5]. Unlike conventional shape alloys [3, 4], the speed of shape change is not limited in this mechanism. Strains are an order of magnitude higher than those observed in rare earth systems, e.g Terfenol D ($Tb_{0.27}Dy_{0.73}Fe_2$), and have been reported in the Ni-Mn-Ga system [5-7]. Ferromagnetic shape memory alloys such as Ni-Mn-Ga [8, 9], Fe-Pd [10], Fe-Pt [11], Co-Ni-Al [12, 13], Co-Ni-Ga [14], Ni-Fe-Ga [15-17] and Ni-Mn-Z (Z=In,Sn,Sb) [18-20] are among the materials currently being investigated. Many of these materials tend to be brittle and the phase change essential for shape memory behaviour occurs well below room temperature thus limiting the potential for applications. Changes in stoichiometry and atomic order can lead to improved ductility and transformations temperatures above 300 K. With the exception of Ni_2MnGa the transformation sequences and associated structures responsible for shape memory behaviour have not been fully established.

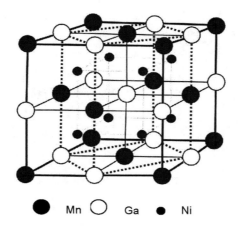

●　Mn　○　Ga　•　Ni

FIGURE 5. The Heusler L2$_1$ austenite structure with the equivalent bct with $a_{Te} = a/\sqrt{2}$ and $c_{Te} = a$ indicated by the broken line.

The Properties of Ni$_2$MnGa

The shape memory effect in the ferromagnetic alloy Ni$_2$MnGa is due to the martensitic phase transition from a high-temperature Heusler structure (Figure 5) to a related structure of lower symmetry at T_M = 200K [8]. Below 200 K the magnetisation, shown in Figure 6, is field-dependent with the full magnetisation only being restored in fields > 0.9T. In higher fields the magnetisation is slightly higher in the martensitic state owing to the change in atomic distances. The moment, which is close to 4 μ_B, is primarily located on the Mn atoms with only ~ 0.3 μ_B associated with the nickel atoms. Above the Curie temperature the uniform susceptibility is Curie–Weiss with a paramagnetic moment $\mu_P = 2S$, with $P^2_{eff} = \mu_P(\mu_P+2)$, very close to the ground state moment μ_{00}. A summary of the bulk properties is given in Table 1 and 2. The Curie temperature of stoichiometric Ni$_2$MnGa is T_C~360K and the structural phase transition which occurs in the ferromagnetic state is not driven by a collapse of the magnetic moment, Figure 6. The change in atomic volume at T_M is < 2%. Band structure calculations [21] suggest that the phase transition at T_M is driven by a band Jahn-Teller mechanism involving a redistribution of electrons between 3d sub-bands of different symmetries and this contention was supported by a polarised neutron study [22] which showed that at T_M there is a transfer of moment from Mn to Ni. Furthermore the results show a re-population of the xz±yz at the expense of the xy t$_{2g}$ orbitals on the Mn atoms and an increased population of the 3z^2-r^2 orbitals on the Ni atoms. However the band structure calculations do not take into account the presence of a premartensitic phase stable between 260 and 200 K associated with an incomplete softening of the TA2 [110] phonon branch at q$_p$ = (⅓ ⅓ 0) [23]. The softening begins above the Curie temperature and does not give rise to an incommensurate 'tweed' phase. A combination of neutron powder [24] and single crystal diffraction [25] measurements have revealed that both the pre-martensitic and martensitic phases have modulated structures which can be described by an orthorhombic unit cell, space

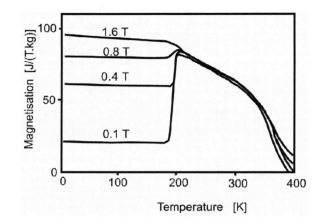

FIGURE 6. The thermal variation of the bulk magnetisation of Ni_2MnGa measured in several applied fields.

group Pnnm, with lattice parameters $a_{ortho} = 1/\sqrt{2}a_{cubic}$, $b_{ortho} = n/\sqrt{2}a_{cubic}$ and $c_{ortho} = a_{cubic}$, $n = 3$ in the premartensitic phase and $n = 7$ in the martensitic phase. A schematic phase diagram is presented in Figure 7. The lattice parameters of the transformed phase are $a = 4.2152$ Å, $b = 29.3016$ Å and $c = 5.5570$ Å indicating that the orthorhombic distortion $(a - b/7) = 0.0293$ Å is small. A combination of polarized neutron and single crystal diffraction measurements have established that the displacements of the Ni, Mn and Ga atoms in both modulated phases of Ni_2MnGa are in phase with one-another and that there are two distinct Mn moments dependent on the Ni-Mn separation. Such correlation has important implications for the control of shape memory properties using a magnetic field.

The martensitic phase transition in Ni_2MnGa, fundamental to its shape memory behaviour, can be described by two successive 110 type shears leading to 36 possible different orientations for the axes of the pseudo-tetragonal martensitic phase. For Ni_2MnGa it was shown [25] that the shears occurring on {110} planes in the <-110> directions lead to tetragonal structure when the shears are equal or when one is twice the other. The general forms of the matrices relating the axes of the tetragonal and cubic cells for these two types when the unit of shear is σ are:

$$\mathbf{M}_{1xy} = \begin{pmatrix} 1+\sigma & 0 & \sigma \\ 0 & 1+\sigma & \sigma \\ -\sigma & -\sigma & 1-2\sigma \end{pmatrix} \qquad \mathbf{M}_{2xy} = \begin{pmatrix} 1+\sigma & 0 & \sigma \\ 0 & 1-2\sigma & -2\sigma \\ -\sigma & 2\sigma & 1+\sigma \end{pmatrix}. \qquad (1)$$

The first index in the subscript indicates whether it arises from equal shears (type 1) or if the second shear is twice the first (type 2). The second and third indices identify the axes perpendicular to the two shears, for type 2 twins the second index corresponds to the longer displacement. For the type 1 domains it is the tetragonal c axis about which there is no rotation, whereas for type 2 domains it is one of the other

TABLE 1. Structural parameters of the cubic $L2_1$ austenite phase of Ni_2MnGa at 295 K and the pseudo tetragonal fct structure of the martensitic phase at 4.2K.

a_{295} [Å]	V_{295} [Å³]	T_M [K]	$a_{4.2}$ [Å]	$c_{4.2}$ [Å]	c/a	$V_{4.2}$ [Å³]
5.825	198	200	5.92	5.566	0.94	195

TABLE 2. A summary of the bulk ferromagnetic properties below T_C and the paramagnetic properties above.

μ_{00} [μ_B]	T_C [K]	Θ_P [K]	P_{eff}/Mn [μ_B]	μ_P/Mn [μ_B]	μ_P/μ_{00}
4.17	376	378	4.75	3.85	0.893

a axes. The shear displacement is $\sigma = 2(a-c)/3(a+c)$. For type 1 shears the mean rotation about the three axes is $2\sigma/3$ whereas for type 2 it is σ. The transformation matrices describing the two periodic displacements of atoms in successive {110} planes along the [1-10] direction gives rise to an invariant [111] direction. It has also been found that two shears are required to produce a coherent interface.

The effect on the structures of applying uniaxial stress or a magnetic field has been studied both above and below T_M. Variants whose c-axes are parallel to the stress are favoured, whilst those with perpendicular c axes, disappear. A very similar behaviour is observed on applying a magnetic field; 0.03 T was sufficient to convert all twins whose c-axes were perpendicular to the field direction to ones with parallel c-axes. These results suggest that magneto-strictive strain rather than magneto-crystalline anisotropy is responsible for the change in domains brought about by applying a magnetic field.

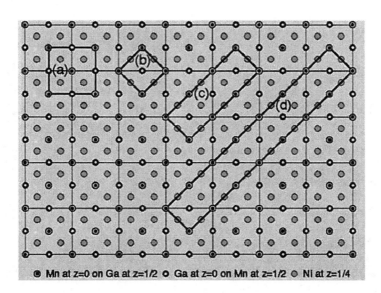

● Mn at z=0 on Ga at z=1/2   ○ Ga at z=0 on Mn at z=1/2   ◎ Ni at z=1/4

FIGURE 7. A projection on to the (001) plane of the different structures observed in Ni_2MnGa indicating their inter-relationships, (a) $L2_1$ structure, (b) bct cell, (c) 3 fold modulated premartensitic phase and (d) the 7 fold modulated martensitic phase.

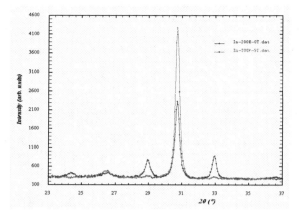

FIGURE 8. A partial neutron powder diffraction pattern of $Ni_{46}Mn_{41}In_{13}$ recorded at 200 K in zero field and with a vertical field of 5 T. It can be seen that the effect of applying a magnetic field is to transform the martensite into the austenite phase.

Although a large strain, up to $(1-c/a) \sim 12\%$ [26], can be obtained for a multi-variant tetragonal martensite the output stress is limited to ~ 4 MPa. This shortcoming may be remedied if the martensitic transformation, essential for shape memory behaviour, is itself field-dependent [27]. The change ΔT in the transformation temperature resulting from a variation ΔB in applied field can be estimated using the Clausius-Clapeyron relation:

$$\Delta T \approx \left(\frac{\Delta\sigma}{\Delta S}\right)\Delta B, \qquad (2)$$

where $\Delta\sigma$ and ΔS are the difference in magnetisation and entropy between the austenite and martensite phases. To produce a large change in ΔT requires a large $\Delta\sigma$ and small ΔS. Owing to the small $\Delta\sigma$ in Ni-Mn-Ga alloys TM varies only by ~ 1.6 K in fields of 2T. Ferromagnetic alloys in the Ni-Fe-Ga series have been reported to have shape memory properties superior to those of other systems [15, 16]. Both T_C and T_M can be controlled either by heat treatment which affects the degree of L2$_1$ order or by changing composition. Moreover, the ductility can be improved by the introduction of the γ phase into the grain boundaries. At the composition $Ni_{54}Fe_{19}Ga_{27}$ T_M and T_C coincide at 296 K. Furthermore single crystal neutron diffraction measurements in a magnetic field show that T_M increases by 0.3 K/T [17]. Changing the stoichiometry in the $Ni_{2+x}Mn_{1-x}Ga$ series causes T_C to decrease and T_M to increase with the two temperatures coinciding at ~ 325 K for x = 0.19 [9]. In a sample in which the two temperatures coincided at 350 K, 30% of the volume was found to be in the austenite phase and 70% in the martensitic phase. However the percentage of cubic phase was reduced to 15 % by applying a field of 5 T [28]. On removal of the field the volume of the cubic phase only returned to 22 % further restricting possible applications. A more significant dependence of T_M has been reported for alloys in the $Ni_{50}Mn_{50-x}In$ series. Magnetisation measurements on $Ni_{46}Mn_{41}In_{13}$ indicate a Curie temperature of 310 K and a structural phase transition T_M at 213 K. In a field of 5.5 T, T_M was found to be reduced to 180 K. Neutron powder diffraction measurements show that the martensitic phase has a complex monoclinic structure and that fields up to 5 T only partially transform the material as shown in Figure 8. However the switching is well below room temperature as required for most applications.

CONCLUSIONS

Although significant progress has been made in identifying possible magnetic shape memory materials as yet there is no ideal system for use in medical applications. Problems such as the brittle nature of some of the materials can be overcome by encapsulation in a polymeric binder. Alloys in the Ni-Co-Ga or Ni-Co-Al systems [12, 13] are ductile and have transition temperatures close to room temperature and in the future may prove of use. Recently Co doped Ni-Mn-In [18] and Ni-Mn-Sn [29] are showing promise for possible applications. However, in most of these systems the structures and transformation mechanisms responsible for shape memory behaviour are unknown and require considerable further investigation.

REFERENCES

1. T. Duerig, A. Pelton and D. Stöckel, *Material Science and Engineering: A* **273-275**, 149 (1999).
2. L.G. Machado and M.A. Savi, *Brazilian Journal of Medical and Biological Research* **36**, 683 (2003).
3. Z. Nishiyama, in *Martensitic Transformation*, edited by M.E. Fine, M. Meshii and C.M. Wayman, London: Academic Press, 1978.
4. K. Otsuka and C.M. Wayman, *Shape Memory Materials*, Cambridge: Cambridge University Press, 1998.
5. K. Ullakko, J.K. Huang, C. Kantner, R.C. O'Handley and V.V. Korkorin, *Appl. Phys. Letts.* **69**, 1966 (1996).
6. S.J. Murray, M. Morioni, S.M. Allen, R.C. O'Handley and T.A. Lograsso, *Appl. Phys. Letts.* **77**, 886 (2000).
7. I. Soursa, E. Pagounis and K. Ullakko, *J. Mag. Magn. Mat.* **272-276**, 2029 (2004).
8. P.J. Webster, K.R.A. Ziebeck, S.L. Town and M.S. Peak, *Phil. Mag.* **49**, 295 (1984).
9. A.N. Vasil'ev, A.D. Bozhko, V.V. Khovailo, I.E. Dikshtein, V.G. Shavrov, V.D. Buchelnikov, M. Matsumoto, S. Suzuki, T. Takagi and J. Tani, *Phys. Rev. B* **59**, 1113 (1999).
10. R.D James and M. Wuttig, *Phil. Mag. A* **77**, 1273 (1998).
11. T. Kakeshita, T. Takeuchi, T. Fukuda, T. Saburi, R. Oshima, S. Muto and K. Kishio, *Mater. Trans. JIM* **41**, 882 (2000).
12. K. Oikawa, L. Wulff, T. Iijima, F. Gejima, T. Ohmori, A Fujita, K. Fukamichi, R. Kainuma and K. Ishida, *Appl. Phys. Letts.* **79**, 3290 (2001).
13. H. Morito, A Fujita, K. Fukamichi, R. Kainuma, K. Ishida and K. Oikawa, *Appl. Phys. Letts.* **81**, 1657 (2002).
14. M. Wuttig, J. Li and C. Craciunesca, *Scr. Mater.* **44**, 2393 (2001).
15. K. Oikawa, T. Ohta, Y. Tanaka, H. Morito, T. Ohmori, A Fujita, R. Kainuma, K. Fukamichi and K. Ishida, *Appl. Phys. Letts.* **81**, 5201 (2002).
16. K. Oikawa, T. Ohta, Y. Sutou, T. Ohmori, R. Kainuma and K. Ishida, *Mater. Trans. JIM* **43**, 2360 (2002).
17. P.J. Brown, A.P. Gandy and K.R.A. Ziebeck, *J. Phys.: Condens. Matter* **18**, 2925 (2006).
18. K.Oikawa, W. Ito, Y. Imano, Y. Sutou, R. Kainuma, K. Ishida, S. Okamoto, O. Kitakami and T. Kanomata, *Appl. Phys. Letts.* **88**, 122507 (2006).

19. Y. Sutou, Y. Imano, N. Koeda, T. Omari, R. Kainuma, K. Ishida and K. Oikawa, *Appl. Phys. Letts.* **85**, 4358 (2004).
20. P.J. Brown, A.P. Gandy, K. Ishida, R. Kainuma, T. Kanomata, K.-U. Neumann, K. Oikawa, B. Ouladdiaf and K.R.A. Ziebeck, *J. Phys.: Condens. Matter* **18**, 2249 (2006).
21. S. Fujii, S. Ishida and S.J. Asano, *J. Phys. Soc. Japan* **58**, 3657 (1987).
22. P.J. Brown, A.Y. Bargawi, J. Crangle, K.-U. Neumann and K.R.A. Ziebeck, *J. Phys. Condens. Matter* **11**, 4715 (1999).
23. A. Zheludev, S.M. Shapiro, P Wochner and L.E. Tanner, *Phys. Rev. B* **54**, 5045 (1996).
24. P.J. Brown, J. Crangle, T. Kanomata, M. Matsumoto, B. Ouladdiaf, K.-U. Neumann and K.R.A. Ziebeck, *J. Phys.: Condens. Matter* **14**, 10159 (2002).
25. P.J. Brown, B. Dennis, J. Crangle, T. Kanomata, M. Matsumoto, K.-U. Neumann, L.M. Justham, K.R.A. Ziebeck, *J. Phys.: Condens. Matter* **16**, 65 (2004).
26. K. Inoue, B. Dennis, T Kanomata, K.-U. Neumann and K.R.A. Ziebeck, *Int. J. of Applied Electromagnetics and Mechanics* **21**, 159 (2005).
27. W. Ito, Y. Imano, R. Kainuma, Y. Sutou, K. Oikawa and K. Ishida, *Metall. and Mat. Trans.A* **37**, 1 (2006).
28. K. Inoue, K. Enami, Y. Yamaguchi, K. Ohoyama, Y. Morii, Y. Matsuoka and K. Inoue, *J. Phys. Soc. Japan* **69**, 3485 (2000).
29. R. Kainuma, Y. Imano, W. Ito, H. Morito, Y. Sutou, K. Oikawa, A. Fujita and K. Ishida, *Appl. Phys. Letts.* **88**, 192513 (2006).

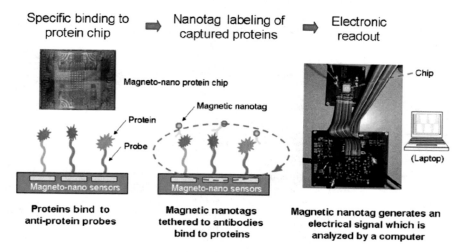

Specific binding to protein chip ➡ Nanotag labeling of captured proteins ➡ Electronic readout

Magneto-nano protein chip

Magnetic nanotag

Protein

Probe

Magneto-nano sensors

Magneto-nano sensors

Chip

(Laptop)

Proteins bind to anti-protein probes

Magnetic nanotags tethered to antibodies bind to proteins

Magnetic nanotag generates an electrical signal which is analyzed by a computer

Detection of proteins using the magneto-nano protein chip. Taken from S.X. Wang, "**Giant Magnetoresistive Biochips for Biomarker Detection and Genotyping: An Overview**".

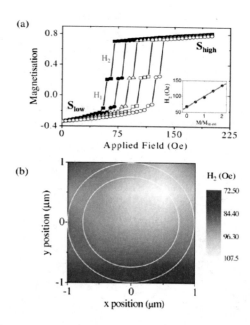

(a) Simulated switching of a PSV ring sensor when shielded by magnetic beads of various saturation moments (M_{bead}). The switching proceeds from the low resistance state (S_{low}) to the high resistance state (S_{high}). Closed squares: $M_{bead}/M_{M-450} = 0$. Closed circles: $M_{bead}/M_{M-450} = 0.5$. Open triangle: $M_{bead}/M_{M450} = 1$. Open squares: $M_{bead}/M_{M-450} = 1.5$. Open circles: $M_{bead}/M_{M-450} = 2$. M_{M-450} represents the saturation moment of an M-450 Dynabead®. The inset figure shows the variation of H_2 with the moment of the bead. (b) Variation of switching field H_2 with the position of a bead with $M_{bead}/M_{M-450}=1$. The white lines indicate the outline of the ring sensor. Taken from T.J. Hayward *et al.*, "**Towards Magnetic Suspension Assay Technology**".

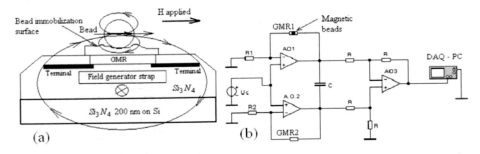

(a) The schematic setup for a GMR sensor that detects the stray field generated by a bead. The field that magnetizes the immobilized bead can be generated by a strap located beneath the sensor. (b) The differential measurement setup used to measure the stray field produced by the magnetic bead. The sensor GMR2 is used as reference. Taken from M. Volmer et al., "**Detection of Magnetic-Based Bio-Molecules Using MR Sensors**".

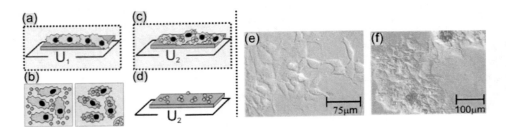

(a-d) Schematic representation of testing: (a)- The voltage drop across the MI sensitive element is measured in presence of solution of cells (b)- The infection where the cell were mixed with the magnetic nanoparticles, which was introduced into the cells by uptake events and the supernatant, free nanoparticles, were removed. (c)- The voltage drop across the MI element installed in a bath which contains the magnetic nanoparticles inside the cell. (d)- The magnetic equivalent of 3 (e, f). Optical image taken before MI measurement of: (e)- HEK 293 cell (f)- HEK 293 cells infected by nanoparticles. Taken from V. Fal-Miyar et al., "**Giant Magnetoimpedance for Biosensing in Drug Delivery**".

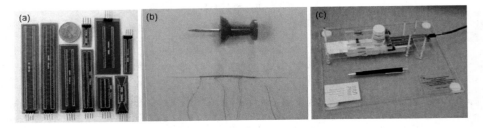

(a) Image of a set of PCB integrated fluxgate sensors. (b) RTD-fluxgate magnetometer in "wire core" technology (microwire-fluxgate magnetometer). (c) The experimental set-up for magnetic beads detection. Taken from C. Trigona et al., "**Residence Times Difference Fluxgate Magnetometer for Magnetic Biosensing**".

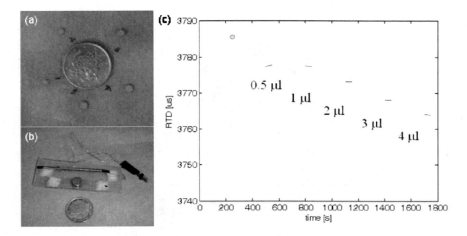

(a) Magnetic beads (CM-10-10) immobilized onto 100 μm-thick glass surface. (b) The experimental set-up for spotted magnetic beads detection. (c) RTD evolution for the microwire fluxgate magnetometer; the graphs reports the experimental results obtained by applying the sensor to target with an increasing number of magnetic particles deposited. Taken from C. Trigona *et al.*, "**Residence Times Difference Fluxgate Magnetometer for Magnetic Biosensing**".

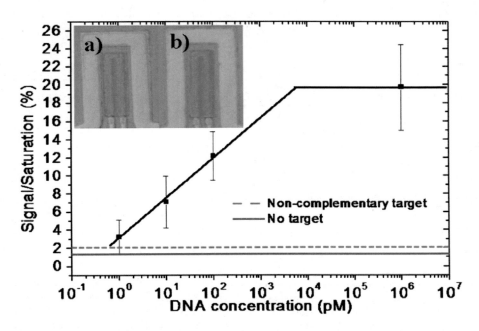

Biological detection limit for single stranded DNA sequences encoding for the genomic region of the 16S ribosomal sub-unit of *E. coli*. The straight lines connecting the experimental data points are simply a guide to the eye. *Inset:* Optical microscope pictures at 800x magnification, from individual sensors corresponding to (a) 1 μM target and (b) no target assay. Taken from V.C. Martins *et al.*, "**Integrated Spintronic Platforms for Biomolecular Recognition Detection**".

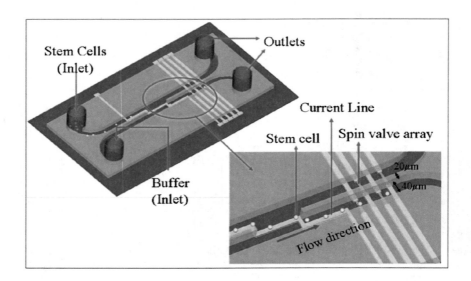

"H-type" fluidic platform allowing magnetically labeled cells to be separated from the left to the right channel due to the magnetic field created from the current line. Each channel has spin-valves at the end to count the cells. Taken from V.C. Martins *et al.*, "**Integrated Spintronic Platforms for Biomolecular Recognition Detection**".

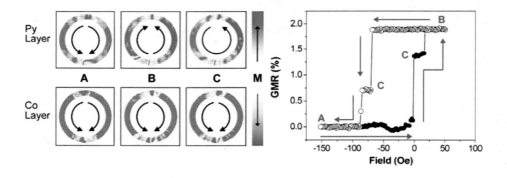

Micromagnetic simulations and measured MR response of PSV ring. Micromagnetic simulations using OOMMF show magnetization states of a 2μm PSV ring with 200nm linewidth as the Py layer is cycled from reverse onion (A) to forward onion state (B) and back, as indicated by the arrows. MR measurements on an actual microfabricated ring confirm that the reverse onion state of the Co layer remains largely undisturbed during the cycling, giving rise to minimum (A) and maximum (B) MR levels. The plateaus correspond to intermediate states (C) obtained in the Py layer during the ascending and descending field sweeps. Taken from J. Llandro *et al.*, "**Moment Selective Digital Detection of Single Magnetic Beads for Multiplexed Bioassays**".

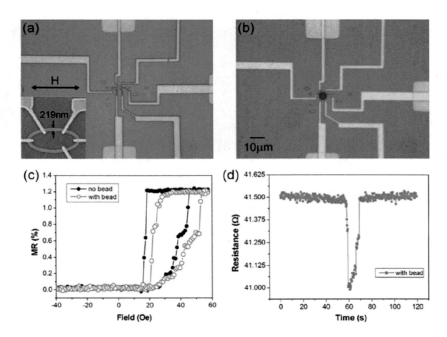

(a) and (b) A single 4.5μm M-450 Dynabead® is positioned over the elliptical PSV ring (4μm long axis, 2μm short axis, 219nm linewidth). Inset shows scanning electron micrograph of the ring, indicating the linewidth and the applied field direction. (c) Minor loop of the free layer of 4μm elliptical PSV ring, showing transitions in absence (closed circles) and presence (open circles) of the bead. (d) Resistance vs. time response of the elliptical ring for a critically balanced field sweep. An abrupt drop in the resistance occurs in the presence of a single Dynabead®. The ring was supplied with a constant current of 40μA. Taken from J. Llandro *et al.*, "**Moment Selective Digital Detection of Single Magnetic Beads for Multiplexed Bioassays**".

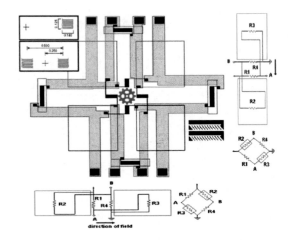

The two-axis independent Wheatstone bridges sensor layout. Taken from M. Avram *et al.*, "**Advanced Magnetoresistance Sensing of Rotation Rate for Biomedical Applications**".

PART 2

MAGNETIC BIOSENSORS AND DETECTION SYSTEMS

Giant Magnetoresistive Biochips for Biomarker Detection and Genotyping: An Overview

Shan X. Wang[a,b]

[a]*Department of Materials Science and Engineering, Stanford University, Stanford, CA 94305-4045, USA.*
[b]*Department of Electrical Engineering, Stanford University, Stanford, CA 94305-4045, USA.*

Abstract. Giant magnetoresistive biochips based on spin valve sensor arrays and magnetic nanoparticle labels have been successfully applied to the detection of biological events in the form of both protein and DNA assays with great speed, sensitivity, selectivity and economy. The technology is highly scalable to deep multiplex detection of biomarkers in a complex disease, and amenable to integration of microfluidics and CMOS electronics for portable applications. The results suggest that a magneto-nano biochip holds great promises in biomedicine, particularly for point of care molecular diagnostics of cancer, infectious diseases, radiation injury, cardiac and other diseases.

Keywords: Spin valve sensor; GMR; Magnetic nanoparticles; Biodetection; Biosensor.
PACS: 78.67.Bf; 87.80.Fe; 87.85.fk

INTRODUCTION

Biosensors and biochips are widely used in biomedical research and practices, prominently among which are microarrays or biochips that typically assay analytes, such as nucleic acids or proteins in a solution via spatially separated molecular probes immobilized on a solid surface. For example, DNA and protein microarrays have played an increasingly important role in gene expression and proteomic studies and other high throughput applications. Most commercial microarray systems utilize fluorescent labeling (tagging) to quantify biomolecular analytes (targets). They are inherently of limited sensitivity because they require approximately 10^4 or more molecules to achieve a useful signal-to-noise ratio and are marginally quantitative because of the optical systems involved, of crosstalk and of bleaching [1]. The optical detection systems are usually used in conjunction with amplification techniques such as polymerase chain reaction (PCR) which multiplies the original biomolecules by many orders of magnitude. Therefore, alternative microarray technologies with a higher sensitivity, lower cost, and better portability are sought after. Such technologies can open many new applications in the field of molecular diagnostics, proteomics, and DNA fingerprinting. It can also form the basis for a rapid response system required in triaging radiation exposure after a radiological incident or genotyping suspected pathogenic agents (e.g. bacteria and virus) in large-scale infectious diseases such as tuberculosis and bioterrorist outbreak such as anthrax. To this end, we at Stanford

CP1025, *Biomagnetism and Magnetic Biosystems Based on Molecular Recognition Processes*
edited by J. A. C. Bland and A. Ionescu
© 2008 American Institute of Physics 978-0-7354-0547-9/08/$23.00

have been developing a magneto-nano biochip [2, 3] based on spin valve (SV) or magnetic tunnel junction (MTJ) sensor arrays and magnetic nanoparticle tags.

A group at the Naval Research laboratory and NVE Corporation in the US first demonstrated a magnetic biosensor system which they called BARC [4-5]. Their detector employed a giant magnetoresistive multilayer stack that is less sensitive than properly designed spin valve (SV) or magnetic tunnel junction (MTJ) detectors [6]. Another group in Germany demonstrated that giant magnetoresistive (GMR) multilayer biodetection was superior to fluorescent biodetection [7]. Groups in Portugal and Netherlands have deployed SV sensors coupled with coils at proximity for rapid molecular detection [8, 9]. The commercially available magnetic tags used by these groups tend to have a mean diameter in the range from 0.1 μm to 3 μm, including the paramagnetic polystyrene beads and similarly sized ferromagnetic particles. The larger tags are seriously mismatched in size with the typical DNA fragments or protein targets in biological assays, prejudicing the quantitative capabilities of the system.

The most tangible and unique feature of the magneto-nano biochip in our work is that it uses magnetic nanoparticles (also called NanoTags or nanotags) with a mean diameter of only 100-1000 Å as the tags. Since the tag dimensions are more comparable to those of the target biomolecules to be assayed, we expect better performance in real biological assays. Another salient feature of our magneto-nano chip is that our SV detectors are typically about one micrometer to submicron in width (but not length), which is about one to two orders of magnitude larger than the dimension of nanotags. Furthermore, the sensor arrays are coated with an ultrathin but corrosion-resistive passivation layer, allowing the biomolecule targets and nanotags to be very close to the magnetic sensing layer, facilitating real-time magnetic readout without washing. Finally, the nanotags are excited by a modulating magnetic field, reducing the effect of 1/f noise and enabling the biological signal readout in a narrow frequency band in spite of noises and interferences. Because of these features, detection of a single nanotag and a detector density of $\sim 10^6$ detectors/cm^2, integrated with CMOS electronics, are feasible on a mass-produced magnetic microarray by using our existing technology.

PRINCIPLE OF GMR BIOSENSOR

The combination of SV (or MTJ) sensor arrays and magnetic nanotags constitutes a promising architecture for a sensitive, quantitative, non-optical detection system for microarrays, a universal platform for many different biological assays. The basic methodology of such a magnetic microarray in the case of DNA detection is shown in Figure 1: (a) SV or MTJ sensor (detector) array is fabricated by optical or e-beam lithography, then bound with known DNA probes. (b) Unknown DNA fragments (targets) are labeled by high moment magnetic nanotags via binding mechanisms such as biotin-streptavidin chemistry. (c) Tagged DNA fragments are selectively captured by complementary DNA probes, and the magnetic nanotags are read out by SV or MTJ sensors. This process is called *direct labeling*. Alternatively, DNA targets can be

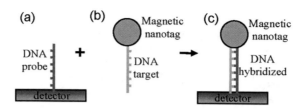

FIGURE 1. Principle of a magneto-nano biosensor with a DNA target as model analyte.

hybridized probes on the chip first, and then be labeled with nanotags. The latter process is called *indirect labeling*.

The magneto-nano biochip is a magnetic biodetection system and its sensors are by nature magnetic field detectors which are sensing the magnetic fringing field of captured magnetic nanoparticle labels. However, the fringing field from magnetic nanoparticles is not uniform across the sensor, and the average fringing field is usually small due to the very limited particle size, the relatively large sensor dimension, and the inverse dependence of the dipole field magnitude on the cube of the distance from the particle to the sensor. The voltage signal generated at the sensor can be as small as ~1 µV, more than two orders of magnitude smaller than those in magnetic recording. Therefore, not only a field sensitive material/structure but also its proper design is essential for magnetic biosensing. The details of the GMR and MTJ sensor design and fabrication from our group are described in references [2, 3, 9-12]. Experimentally, we previously showed that the limit of detection (LOD) of a spin valve sensor (0.2 µm wide, 1.5 µm long), even without using lock-in (narrow band) detection, is ~14 Fe_3O_4 nanoparticles (each with a mean diameter of 16 nm) [10]. This LOD corresponds to a total magnetic moment of ~11 femto-emu, much smaller than the magnetic moments of most commercial magnetic beads except for those used for magnetic-activated cell sorting (MACS®).

GMR BIOCHIPS FOR BIOMARKER DETECTION AND MOLECULAR DIAGNOSTICS

Molecular diagnosis is a relatively new way of identifying a disease by studying molecules, such as proteins, DNA, and RNA, in a tissue, such as biopsy or in a body fluid such as blood and urine [13]. A molecule or molecules indicative of the presence or state of a particular disease are often called biomarkers specific to the disease. More broadly, a biomarker can be defined as a "characteristic that is objectively measured and evaluated as an indicator of normal biological processes, pathogenic events, or pharmacologic responses to a therapeutic intervention" [14]. For example, prostate-specific antigen, better known as PSA, is routinely measured in a clinical test or "assay" for prostate cancer screening [15]. If the PSA concentration is high or increasing, a doctor may perform a biopsy to remove tissue samples from the prostate. A pathologist will then examine these samples, looking at the shapes of individual cells and the patterns they form under a microscope, to make a specific diagnosis. Other examples of well known biomarkers include cardiac troponin T (cTnT, a regulatory protein found in cardiac muscle) for myocardial infarction (heart attack) [16], phosphorylation of a special form of histone protein (γ-H2AX) in cells for ionizing radiation exposure [17], and mutation of BRCA1 and BRCA2 genes associated with increased risk of breast and ovarian cancer. Complex diseases like cancer are multifactorial and often involve multiple biomarkers and environmental/behavioral factors. There is a consensus that less invasive and faster molecular diagnosis based on measuring a panel of perhaps 10-20 biomarkers which are specific and predictive of a complex disease will be highly desirable for medical diagnosis and has the potential to revolutionize modern healthcare.

For magnetic biosensors and biochips to be truly useful in molecular diagnostics, they must be sufficiently sensitive as well as selective to detect the biomarkers relevant to a disease of interest. "Sensitivity" is defined as the smallest amount of the target molecule that an assay can detect, while "selectivity" refers to how well an assay can detect particular molecules in a complex mixture without interference from other molecules in the mixture[1]. Most assays are not highly selective because of the cross reactivity between many molecules in the assay. In addition, many existing assays often rely on optical labels that produce or emit light when excited, and most body fluids such as blood contain a host of other compounds that behave similarly. Even worse, the intensity of the emitted light often varies with sample pH and decreases over time, a result of a chemical process known as photobleaching. Magnetic biosensors using magnetic nanotags labeling of biomarkers are ideally suited to avoid some of these problems because all other components in a blood sample solution are essentially non-magnetic and therefore interference effects and background signals are much reduced. Furthermore, the magnetic properties of nanotags are stable over time. With a robust biochemistry, the high "magnetic sensitivity" of small number of nanotags can be translated into the high sensitivity of small number of biomarkers.

As an example, we have designed and fabricated a tiny and inexpensive magneto-nano protein chip (Figure 2) that could correlate the number of magnetic nanoparticle labels on a surface with a voltage signal of a GMR spin valve sensor at the site, while ensuring selective protein binding at multiplex sites. Each individual sensor responds to stray magnetic fields by changing its electrical resistance. As more and more nanotags deposit, their stray magnetic fields cause the electrical resistance of the sensor to decrease in proportion to the number present. There are three basic steps for protein quantitation using the magneto-nano protein chip. First, probes specifically "capture" proteins from the sample and bind them to the sensor surface. Second, nanotag-labeled antibodies bind to these surface-bound proteins. This process is akin to the *indirect labeling* mentioned in Figure 1, and is called sandwich immunoassay. Finally, an external modulating magnetic field is applied to the chip and the stray magnetic field produced by the nanotag labels is measured. The lower the resistance,

---

[1] In the medical profession, sensitivity is defined as TP/ (TP+FN), while selectivity or specificity is defined as TN/ (TN+FP), where TP = true positive, FN = false negative, TN = true negative and FP = false positive.

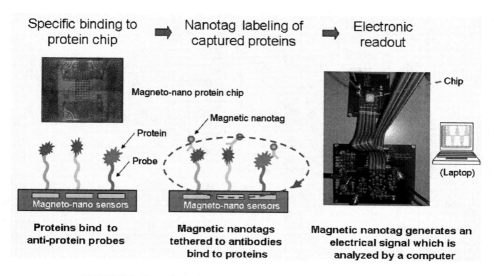

Specific binding to protein chip → Nanotag labeling of captured proteins → Electronic readout

Magneto-nano protein chip

Magnetic nanotag

Protein

Probe

Magneto-nano sensors

Magneto-nano sensors

Chip

(Laptop)

Proteins bind to anti-protein probes

Magnetic nanotags tethered to antibodies bind to proteins

Magnetic nanotag generates an electrical signal which is analyzed by a computer

FIGURE 2. Detection of proteins using the magneto-nano protein chip.

the more nanotags are present and the higher the concentration of protein in the original sample.

Recently, we have prototyped magneto-nano chips (Figure 3) with an 8 x 8 array of 64 giant magnetoresistance (GMR) spin valve sensors, each of which consists of 32 giant magnetoresistance (GMR) strips which are 1.5 µm wide by 110 µm long and in series connection. Each sensor is about 110 µm by 110 µm in area and covered with a unique protein feature which can be spotted with a robotic inkjet or other types of pins. The total area of the magneto-nano chip is about 10 mm by 12 mm, while the active 8 x 8 sensor array (denoted by the dashed square) occupies an area of only 3 x 3 mm^2. The chip is also compatible with our microfluidic sample delivery/washing channels and electronics. The fabricated chips are intrinsically multiplex by virtue of having 64 capture probe spots, enabling multiplex detection of up to 64 biomarkers in one test. Moreover, the sensors under an ultrathin passivation layer have proven to be chemically stable in aqueous solutions or blood samples. These chips are ideal for measuring multiple protein levels in a sample volume of only 10-50 µL of blood, taken e.g. from a human patient with minimal invasiveness in an emergency room or for point of care testing.

In order to show the capability to quantify secreted protein markers, we performed many protein assays with both buffer solution and human blood samples using our prototype magneto-nano chip. For example, the sensor arrays in Figure 3 above were coated with specific antibody probes such as interferon-γ (IFN-γ) antibodies, and then purified IFN-γ antigen solutions or serum samples with elevated IFN-γ levels were incubated with the chip, followed by a detection agent consisting of secondary antibodies tethered to magnetic nanotags. Captured nanotags were counted electronically under a modest AC magnetic field in real time. The net signal is

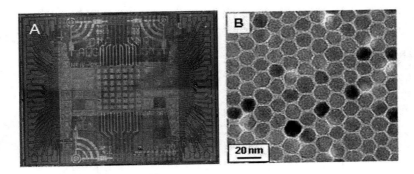

FIGURE 3. (a) Left: Top view of a magneto-nano chip for protein detection. The dashed rectangle indicates the 8 x 8 sensor array. The structures above and below the rectangle are microfluidic channels for serum sample inlet and outlet. The sensors are read out via horizontal wires and bonding pads at the left and right edges of the chip. (b) Right: Representative transmission electron microscopy (TEM) image of monodisperse Fe_3O_4 nanotags. In many of our bioassay experiments we also use commercial MACS® nanotags because the latter are more readily available.

proportional to the number of antibody/IFN-γ/antibody "sandwiches" formed during the assay. Trial runs of our prototype protein chip to detect IFN-γ antigen target are shown in Figure 4, which plots the electronic signal versus time at model analyte concentrations ranging from 0.59 nM to 59 nM. Streptavidin-coated magnetic nanoparticles are applied to the chip at time t = 0, and begin to accumulate on the surface of the spin valve sensors, presumably in proportion to the local surface concentration of captured IFN-γ. Excellent signal-to-noise ratio, selectivity, and fast readout (<30 minutes) were demonstrated. Note that this result is obtained with 1.5 μm wide sensors and MACS® tags. More importantly, the sensor gives the similar signal-to-noise ratios for a given concentration of IFN-γ in blood as in buffer, without the need for any pre-purification. Our latest results showed detection of human chorionic gonadotropin (hCG, a pregnancy marker) protein levels as low as 5 pM (0.2 ng/mL) in blood samples, further confirming the viability of magneto-nano chips for use with real world samples.

GMR DNA CHIP FOR GENOTYPING

Genotyping of pathogens is an important niche of molecular diagnosis, particularly in infectious diseases and biodefense. We have successfully shown our magneto-nano chip can be used for 4-plex Human Papillomavirus (HPV) genotyping [19]. The target samples we used were polymerase chain reaction (PCR) products amplified from cloned plasmid HPV DNA. HPV is a subset of Papillomaviruses that infects the epithelial cells of the skin and mucus membranes in humans. Infection with HPV is associated with various forms of cancers, including cervical cancer. Based on the oncogenic potential, HPV genotypes are classified as "high-risk" (e.g., HPV16, HPV18 and HPV39) or "low-risk" subtypes (e.g., HPV6 and HPV11), so genotyping of the HPV subtypes or strains in a given patient is beneficial to diagnosis and prognosis of cervical cancer. Figure 5 shows the results of an experiment displaying

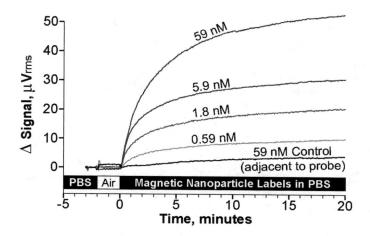

FIGURE 4. Five real-time signals from four different chips. The upper four signal traces were obtained from identically functionalized chips which were subsequently exposed to different concentrations of IFN-γ (antigen) in phosphate buffered saline (PBS) buffer. The control tracing comes from a control sensor that was located inside the same well as the probe sensor which provided the 59 nM signal trace [18].

real-time responses of 3 DNA fragments (Globin A, Globin B and HPV39) at 10 pM applied to a GMR DNA chip identical to those used for IFN-γ assay except that the surface chemistry and capture probes correspond to Globin A, Globin B and HPV39. Note that specific nanoparticle adsorption reaches saturation in mere 10 minutes, suggesting these GMR biochips can be quite fast.

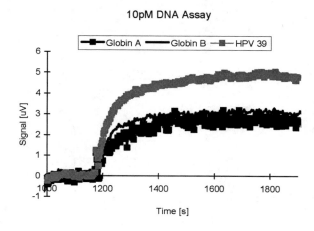

FIGURE 5. Real-time responses of 3 DNA fragments (Globin A, Globin B and HPV39) at concentration of 10 pM applied to a GMR DNA chip.

HIGHLY SCALABLE GMR BIOCHIPS: CMOS INTEGRATED WITH HIGH DENSITY SPIN VALVE SENSOR ARRAYS

To demonstrate the scalability of GMR biochips, we have also designed and fabricated a special version of the magneto-nano chip for molecular diagnostics with high density spin valve sensor arrays fully integrated with CMOS circuitry, with each sensor being 0.3 μm by 5 μm in size [12, 20]. An application specific integrated circuit (ASIC) with a footprint of 2 mm by 2 mm and including row and column addressing decoders and parallel fast readout schemes has been realized. The biochip (Figure 6) features redundant and high density of sensors (16 subarrays per die, 7×9 sensors per subarray), with a sensor density as high as 0.1 million sensors per squared cm. An advanced electronic test station has been set up as a demonstration vehicle for the integrated evaluation of the magnetic biochips with the custom magnetic nanotags and DNA-based biochemistry. The system was shown to detect ~100 nanotags per sensor in DNA hybridization assays [21]. This work suggests that GMR biochips are highly scalable, potentially rivaling the number of biomolecular features in commercial fluorescent microarrays if we can spot distinct capture probes at each sensor.

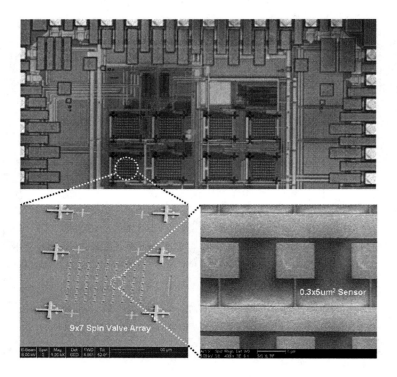

FIGURE 6. Micrograph of the post-processed die and SEM images of the biosensor array. The prototype chip was implemented in a 0.25 μm BiCMOS process (chosen for its availability). Post-processing was performed at the Stanford Nanofabrication Facility (SNF).

ACKNOWLEDGMENTS

The author wishes to acknowledge his deepest appreciation to the late Prof. Tony Bland for organizing the ESF-EMBO Conference "Biomagnetism and Magnetic Biosystems", 22-27 September 2007, St Feliu de Guixols, Spain, and for the privilege of being an invited speaker in this conference. This overview would not have been possible without the contributions from many of the author's collaborators, students, and postdocs, including Robert L. White, Nader Pourmand, Shouheng Sun, Sebastian Osterfeld, Heng Yu, Liang Xu, Shu-Jen Han, Guanxiong Li, Richard Gaster, Drew Hall, Aihua Fu, David Robinson, Robert Wilson, Michael Akhras, Ronald Davis, Kitch Wilson, Mei Huang, Joseph Wu, and Sam Gambhir. This work was supported by Defense Advanced Research Projects Agency (DARPA) Grant N000140210807, National Science Foundation Grant DBI-0551990, National Institutes of Health Grant P01-HG000205, Defense Threat Reduction Agency Grant HDTRA1-07-1-0030, National Cancer Institute Grant 1U54CA119367-01, Semiconductor Research Corp. Grant 2007-RJ-1654G, and National Academies Keck Future Initiatives Award.

REFERENCES

1. M. Schena and R.W. Davis, "Technology Standards for Microarray Research" in *Microarray Biochip Technology*, edited by M. Schena, Westborough, MA: Eaton Publishing, 2000, pp. 1-18.
2. G. Li, V. Joshi, R.L. White, S.X. Wang, J.T. Kemp, C. Webb and R.W. Davis, "Detection of single micron-sized magnetic bead and magnetic nanoparticles using spin valve sensors for biological applications", *J. Appl. Phys.* **93**, 7557-9 (2003).
3. S.X. Wang, S.-Y. Bae, G. Li, S. Sun, R.L. White, J.T. Kemp and C.D. Webb, "Towards a magnetic microarray for sensitive diagnostics", *J. Mag. Mag. Mat.* **293**, 731-6 (2005).
4. D.R. Baselt, G.U. Lee, M. Natesan, S.W. Metzger, P.E. Sheehan and R.J. Colton, "A biosensor based on magnetoresistance technology", *Biosen. Bioelectron.* **13**, 731-9 (1998).
5. J.C. Rife, M.M. Miller, P.E. Sheehan, C.R. Tamanaha, M. Tondra and L.J. Whitman, "Design and performance of GMR sensors for the detection of magnetic microbeads in biosensors", *Sens. Actuator A:Physical* **107**, 209-218 (2003).
6. S.X. Wang and A.M. Taratorin, *Magnetic Information Storage Technology*, San Diego, CA: Academic Press, 1999.
7. J. Schotter, P.B. Kamp, A. Becker, A. Puhler, G. Reiss and H. Brückl, "Comparison of a prototype magnetoresistive biosensor to standard fluorescent DNA detection", *Biosensors and Bioelectronics* **19** (10), 1149 -1156 (2004).
8. D.L. Graham, H. Ferreira, J. Bernardo, P.P. Freitas and J.M.S. Cabral, "Single magnetic microsphere placement and detection on-chip using current line designs with integrated spin valve sensors: biotechnological applications", *J. Appl. Phys.* **91**, 7786-8 (2002).
9. A. Sandhu, "Biosensing: New probes offer much faster results", *Nature Nanotechnology* **2**, News and Views, 748-9, December (2007).
10. G. Li, S. Sun, D.B. Robinson, R.J. Wilson, R.L. White, N. Pourmand and S.X. Wang, "Spin valve sensors for detection of superparamagnetic nanoparticles for biological applications", *Sensors and Actuators A: Physical* **126** (1), 98-106 (2006).
11. S.J. Osterfeld and S.X. Wang, "MagArray biochips for protein and DNA detection with magnetic nanotags: design, experiment, and signal-to-noise ratio", book chapter to appear in *Microarrays: New Development Towards Recognition of Nucleic Acid and Protein Signatures*, edited by K. Dill, R. Liu and P. Grodzinski, Berlin: Springer Verlag/Kluwer, 2008.
12. S.-J. Han, H. Yu, L. Xu, R.J. Wilson, N. Pourmand and S.X. Wang, "CMOS integrated DNA microarray based on GMR sensors", *IEEE International Electron Device Meeting (IEDM) Tech. Dig. Papers*, San Francisco, CA, USA, December 11-13 (2006).

13. http://www.cancer.gov/cancertopics/understandingcancer/moleculardiagnostics
14. F. Marchetti, M.A.Coleman, I.M. Jones and A.J. Wyrobek, "Candidate protein biodosimeters of human exposure to ionizing radiation", *Int. J. Radiat. Biol.* **82** (9), 605-639 (2006).
15. S. Fuessel, D. Sickert, A. Meye, U. Klenk, U. Schmidt, M. Schmitz, A.K. Rost, B. Weigle, A. Kiessling and M.P. Wirth, "Multiple tumor marker analyses (PSA, hK2, PSCA, trp-p8) in primary prostate cancers using quantitative RT-PCR", *International J. of Oncology* **23**, 221-8 (2003).
16. E.M. Antman, M.J. Tanasijevic, B. Thompson, M. Schactman, C.H. McCabe, C.P. Cannon, G.A. Fischer, A.Y. Fung, C. Thompson, D. Wybenga and E. Braunwald, "Cardiac-specific troponin I levels to predict the risk of mortality in patients with acute coronary syndromes", *N. Engl. J. Med.* **335**, 1342-9 (1996).
17. K. Rothkamm and M. Lobrich, "Evidence for a lack of DNA double-strand break repair in human cells exposed to very low x-ray doses", *Proc. Natl. Acad. Sci. USA* **100** (9), 5057-5062 (2003).
18. S.J. Osterfeld, H. Yu et al., "A biochip for rapid and portable protein assays based on magnetoresistive detection of nanoparticles", manuscript in preparation.
19. L. Xu, H. Yu et al., "Giant magnetoresistive biochip for DNA detection and HPV genotyping", submitted to *Biosensors and Bioelectronics* (2008).
20. S.-J. Han, H. Yu, B. Murmann, N. Pourmand and S.X. Wang, "A high density magnetoresistive biosensor array with drift compensation mechanism", *IEEE International Solid-State Circuits Conference (ISSCC) Tech. Dig. Papers*, San Francisco, CA, USA, February 11-15 (2007).
21. S.-J. Han, "CMOS Integrated Biosensor Array Based on Spin Valve Devices", Ph.D. Thesis, Stanford University, CA, USA, September 2007.

Towards Magnetic Suspension Assay Technology

T.J. Hayward[a], J. Llandro[a], K.P. Kopper[a], T. Trypiniotis[a], T. Mitrelias[a], J.A.C. Bland[a] and C.H.W. Barnes[a]

[a]Cavendish Laboratory, University of Cambridge, Cambridge, UK.

Abstract. In this study micromagnetic simulations are used to evaluate two novel approaches of magnetically tagging biomolecules in high-throughput biological assays. Comparisons are made between a simple magnetic moment-based tagging system, where the total magnetic moment of each microscopic tag encodes the identity of an attached biomolecule, and a multibit tagging system, where each tag is comprised of multiple magnetic binary bits. We show that although detection of the tags using magnetoresistive sensors is feasible in both cases, the multibit technology offers over a thousand times more distinct tags than the simple moment encoded approach. The advantages of using multibit magnetic tags to label biomolecules, rather than existing optical tagging techniques, are also discussed.

Keywords: Biosensors, Magnetoresistive sensors, Biochips Magnetic Bead, Magnetic labels, Assay Technology.
PACS: 85.90.+h, 87.85.M- , 87.85.fk

INTRODUCTION

Multiplexed biological assays have become an extremely important tool in contemporary medical research, diagnostics and drug development. Currently the most advanced approach is microarray technology, where target molecules are identified by the observation of a spatially resolved optical signal, produced by the selective binding of fluorescently labelled targets to the complementary probe molecules. State of the art microarray systems are capable of accommodating as many as 1.6 million probe spots.

Although microarray technology is well developed it suffers from several limitations. For example, the identity of probe molecules is encoded by an x-y coordinate position on a 2D plane, and hence binding reactions must occur at a planar surface. Because of this, extra time must be allowed for chemical reactions to occur compared to that which could be achieved with 3D solution-based assays, where solution phase kinetic conditions exist. Furthermore, each probe area in the microarray must be prepared individually either by using a spotting robot or photolithography. Because of this microarrays are relatively expensive to prepare and are difficult to modify if new probes must be incorporated into the assay.

Suspension assay technology (SAT) offers an alternative approach which overcomes some of these limitations [1]. In SAT the binding assay takes place at the surfaces of micron-scale particles which are held in suspension. Each of these particles also carries a characteristic (usually optical) signature which allows the identity of its surface receptor to be determined [2-10]. By incorporating several subsets of particles with distinguishable signatures and different surface receptors into the assay it is possible to screen a sample for several different entities simultaneously, and hence perform a multiplexed bioassay.

In SAT reactions take place in the solution phase and hence can proceed more quickly than is possible at a planar surface. SAT is also much more flexible than microarray technology, as the probe set can be altered simply by adding new sets of particles or replacing existing probes. To modify a microarray in the same manner would require the fabrication of a new chip in its entirety. Furthermore SAT is cheaper than microarray technology, as the encoded particles can be fabricated in bulk, whereas each microarray chip must be assembled at a microscopic level.

Recently there has been substantial interest in the possibility of using magnetic microparticles to replace conventional fluorescent labels in biological assays. Assays are performed by using target molecules to mediate binding between a magnetic bead and the surface of a high sensitivity magnetic field sensor that is functionalised with surface receptors which are complementary to the target molecule [11, 12]. Such an approach is particularly attractive because magnetic particles may be cheaply produced, are very stable with respect to time, resilient to temperature changes and are unaffected by reagent chemistry or exposure to light, unlike conventional fluorescent labels. Furthermore, biological samples, being mostly composed of diamagnetic molecules, have an extremely low magnetic background.

Current methods of using magnetic labels to perform biological assays conceptually resemble microarray technology in that multiplexing occurs by distributing different probe biomolecules across a 2D array of magnetic sensors [13, 14]. These techniques therefore suffer from many of the same limitations in throughput and flexibility associated with microarrays. Because of this, it is important to consider whether a magnetic equivalent of SAT could be created, i.e. where the identity of a particles surface receptors is identified using a magnetic, rather than optical, signature.

In this paper we will use micromagnetic modelling to investigate the feasibility of two different approaches to magnetic tagging. In the first approach, the identity of the tag's surface receptors is encoded by varying the saturation moment of a population of superparamagnetic microspheres (Figure 1(a)). In the second approach the tags consist of a series of thin film rectangular magnetic elements which are supported by a polymer substrate (Figure 1(b)). Each of the magnetic elements exhibits two stable magnetic states which can be used to write a binary code along the tag and therefore encode the identity of the tag's surface receptors.

It will be shown that while encoding using the magnetic saturation moment of a label is possible, it would limit a user's ability to multiplex different assays. In contrast to this, in the second approach, where the labels carry spatially encoded information, several thousand different assays could be performed simultaneously.

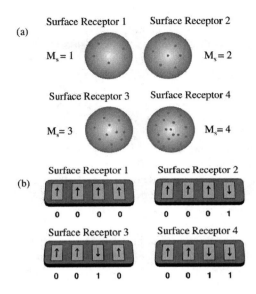

FIGURE 1. Schematic diagrams showing two methods of magnetically tagging biomolecules. (a) Simple magnetic moment-based encoding: Superparamagnetic microspheres are loaded with different amounts of magnetic nanoparticles depending on their surface functionalisation. (b) Planar magnetic tags: The labels consist of a polymer substrate which supports a series of thin film magnetic elements. The magnetisation states of these elements form a binary code which identifies the label's surface functionalisation.

ENCODING USING THE MAGNITUDE OF A LABEL'S MAGNETIC MOMENT

For a magnetic form of SAT to be successful it must allow rapid serial analysis so as to compensate for a reduced degree of parallelism relative to micro-array technology [1]. The most appealing way of implementing this is to analyse the labels in microfluidic flow, as occurs in a conventional flow cytometer. While it is well known that fluorescence techniques can be used to identify binding within a microfluidic system [3, 10], a new method must be devised to rapidly and reliably read the magnetic signature of the labels. Magnetoresistive (MR) stray field sensors are ideally suited to such a task as they are highly sensitive, respond rapidly and can be manufactured using standard lithographic techniques.

To sense the magnetic moments of micron-sized magnetic beads we propose a digital sensing technique which we have previously used to detect magnetic labels with extremely high signal-to-noise ratios [15]. In this approach the soft magnetic layer of a ring-shaped giant magnetoresistive (GMR) sensor [16, 17] is rapidly cycled between a high resistance state and a low resistance state using an external magnetic field. The switching of multilayer ring devices is extremely abrupt [18] and hence the resistance of the sensor alternates almost digitally. If a superparamagnetic bead is introduced above the ring sensor it will be polarised by the external field and produce a magnetic field which will be dominantly in opposition to the external field (Figure 2(a)). The bead therefore screens the sensor so that a larger external field must be applied to switch it between states. This effect can be observed in Figure 2(b) which shows a comparison of magnetoresistance loops measured from a 4 μm GMR ring sensor with and without a superparamagnetic M-450 Dynabead placed in contact with the centre of the ring. As the screening effect is expected to be stronger when a higher moment bead is present, it should be possible to determine the moment of a bead by measuring the offset of the sensor's switching field from its nominal value. To investigate this effect, micromagnetic simulations were performed using the OOMMF software [19] package, in which the soft layer of pseudo-spin-valve (PSV) ring sensor was switched between a low resistance magnetic state (S_{low}) and a high resistance magnetic state (S_{high}) in the presence of superparamagnetic beads with various saturation magnetic moments. The ring had a diameter of 2 μm, a width of 200 nm and a Co(8nm) / Cu(4nm) / Py(4nm) layer configuration. In order to shorten the simulation time only the Py layer (soft layer) of the device was simulated. The magnetostatic field produced by the Co layer (hard layer) was accounted for by introducing a constant anisotropic magnetic field which was extracted from a 3D simulation of the same PSV ring device.

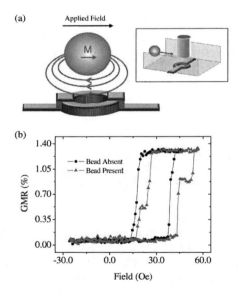

FIGURE 2. (a) Schematic diagram showing a magnetic bead screening a ring shaped biosensor from the influence of an applied magnetic field. The inset figure shows how the beads could be used as tags in high-throughput biological assays: The beads are flowing through a microfluidic channel containing a magnetoresistive sensor and observed by a fluorescence microscope. The sensor allows the identity of the beads surface receptors to be determined, while the fluorescence microscope allows the presence or absence of a target molecule to be detected using conventional fluorescent labelling techniques. (b) Minor MR loops measured from a NiFe/Cu/Co pseudo-spin-valve ring biosensor. Circles: Minor MR loop of the isolated device. Triangles: Minor MR loop when an M-450 Dynabead is placed in contact with the centre of the ring.

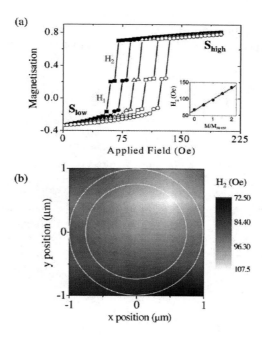

(a)

(b)

FIGURE 3. (a) Simulated switching of a PSV ring sensor when shielded by magnetic beads of various saturation moments (M_{bead}). The switching proceeds from the low resistance state (S_{low}) to the high resistance state (S_{high}). Closed squares: $M_{bead}/M_{M-450} = 0$. Closed circles: $M_{bead}/M_{M-450} = 0.5$. Open triangle: $M_{bead}/M_{M-450} = 1$. Open squares: $M_{bead}/M_{M-450} = 1.5$. Open circles: $M_{bead}/M_{M-450} = 2$. M_{M-450} represents the saturation moment of an M-450 Dynabead. The inset figure shows the variation of H_2 with the moment of the bead. (b) Variation of switching field H_2 with the position of a bead with $M_{bead}/M_{M-450}=1$. The white lines indicate the outline of the ring sensor.

The screening field produced by the magnetic beads was approximated by a dipole field centred a distance R above the ring, where R was the bead radius of 2.2 µm. This represents an ideal case where the beads roll along the channel bottom, minimising the bead-sensor separation.

The magnetic moment of the bead in each field step was calculated from the Langevin function, which was fitted to the M-H curve of a sample of M-450 Dynabeads, as presented in [20]. It was assumed that increasing the magnetic content of the beads produced a linear scaling of the M-H curve. This is unlikely to be true for real superparamagnetic beads where interactions between iron oxide nanoparticles are likely to affect the energy barriers involved in the nanoparticles' magnetic switching, but offers a convenient way of introducing a realistic range of magnetic susceptibilities into the simulation.

Figure 3(a) presents simulated M-H curves for the switching of the Py layer from state S_{low} to state S_{high}. In the absence of the bead the ring switches by a two step mechanism with the first transition (H_1) occurring at 57.5 Oe and the second transition (H_2) at 67.5 Oe. These two transitions represent the magnetic switching of the two sides of the ring. When a magnetic bead with a saturation moment equal to that of an

M-450 Dynabead is introduced above the ring, H_1 and H_2 increase to 87.5 Oe and 97.5 Oe respectively, thus demonstrating that the shielding effect observed experimentally is reproduced in the simulation. Further simulations performed in the presence of beads with saturation moments between half and twice that of an M-450 Dynabead showed that the ring's switching fields varied linearly with magnetic content as can be observed for transition H_2 in the inset to Figure 3(a).

The results described above demonstrate that it should be possible to use a GMR ring sensor to differentiate magnetic microbeads with different magnetic moments. However they do not reveal how many unique tags would be available in such an encoding system. To calculate this, three quantities must be known: the minimum obtainable bead moment, the maximum obtainable bead moment and the range of switching field offsets produced by a subset of beads with nominally identical magnetic moments. The first two factors determine the total range of switching fields that the sensor will exhibit in response to the magnetic beads and therefore the total phase space which can be exploited to encode the tags. The final factor reflects the fact that due to slight variations in the system parameters, such as the beads position as it is sensed, each subset of beads is likely to produce a distribution of switching fields rather than a single discrete value. For a subset of beads to be identified unambiguously, there must be minimal overlap between the distribution of switching fields they produce, and those produced by the subsets of beads with the next highest and next lowest magnetic moments. The number of distributions that can be accommodated in the available switching field phase space determines how many unique tags can be provided.

The minimum available bead moment was taken to be that of a bead with negligible magnetic content. The maximum obtainable moment will depend on the process used to fabricate beads, but in this study was represented by beads containing 50% iron oxide with respect to mass (approximately twice the magnetic content of an M-450 Dynabead). This is close to the maximum magnetic content that can be achieved using a standard fabrication process [21].

The sensor switching field distributions produced by each subset of magnetic beads were represented by Gaussian probability distributions. The widths of these distributions were derived by considering the various factors which could cause variation in the sensor switching field when measuring a particular subset of beads. The first factor considered was the intrinsic variation in the sensor's switching field, which occurs even in the absence of the beads screening field, due to thermal fluctuations and the magnetisation history of the ring. In PSV ring structures the switching field variability typically has a standard deviation (σ) of 1.5 Oe [18].

Figure 4(a) shows the sensor switching field distributions calculated by including only the intrinsic switching field variation of the sensor. Each peak in the plot represents a subset of beads with a different mean saturation moment. The mean values of the distributions were separated from each other by 4σ so that the identity of a tag could be determined with 95% certainty. It can be seen that eleven different subsets of beads can be accommodated between the minimum and maximum switching field values.

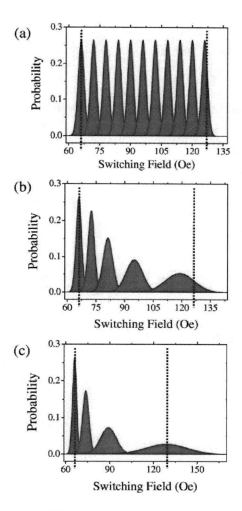

FIGURE 4. Switching field probability distributions for a ring sensor detecting populations of beads with different magnetic moments. Each peak represents a subset of beads with a different mean moment. The dotted lines indicate the maximum and minimum switching fields that can be obtained within the model. (a) Distribution widths reflect the intrinsic variation of the sensors switching field. (b) Distribution widths also include variation in the magnetic content of the beads. (c) Distribution widths allow for uncertainty in the beads position in addition to the previous two factors.

A second factor which would contribute to the widths of the switching distributions is the deviation of the bead's saturation moments from their intended value. In our model this was represented by introducing a variability of the beads moments characterised by a σ of 14% of the mean value. It should be noted that this is a rather conservative estimate of the distribution of moments currently found in commercial superparamagnetic beads [22].

Figure 4(b) illustrates how the sensor switching distributions are broadened by allowing for variability in the moments of the beds in addition to the intrinsic switching variability of the sensor. For beads with low magnetic moments, the intrinsic variability of the sensor still dominates, however for beads with higher moments the distributions are substantially broadened. Consequently, when considering both forms of variability simultaneously, only five subsets of beads can be accommodated between the minimum and maximum switching fields.

In the device we propose the magnetic tags are to be read while in microfluidic flow. The implication of this is that there will be some uncertainty in the position of the tag as it is measured. To investigate the effect of this we performed simulations with the bead placed in different positions within the plane of the sensor. The results of this study are presented graphically in Figure 3(b), where it can be seen that the device is most sensitive to beads positioned within its top right quadrant. For the purposes of the model it was assumed that it was possible to sense the beads as they passed through this region.

Figure 4(c) illustrates how the sensor switching distributions are altered when uncertainty in the bead position is allowed for, in addition to the sensor's intrinsic switching field variability and the variability in the magnetic moment of the beads. The distributions are again wider and only four different bead subsets can be accommodated in the available phase space.

The above calculation has been performed for an example system and therefore does not represent the ultimate limit of what could be achieved using a magnetic moment-based encoding system. However, it is clear that such an approach is likely to provide a rather limited number of unique tags, even with substantial optimisation. It should also be noted that a real system would involve even more degrees of freedom, such as the fly-height of the tags, which would further widen the distributions for a given bead subset and therefore decrease the number of bead subsets which could be successfully discriminated. An alternative approach to creating magnetic SAT is therefore desirable.

PLANAR MAGNETIC TAGS

The simulations described above indicate that a magnetic moment-based tagging system is rather inefficient. Practical difficulties aside, the main reason for this is that for every additional assay that is to be multiplexed we must add a subset of beads with a magnetic moment that can be unambiguously identified. This represents a linear relationship between the number of available codes and the number of distinguishable entities. Binary codes represent a much more efficient method of storing information because a tag composed of N binary bits can accommodate 2^N unique codes. There is therefore an exponential relationship between the number of distinguishable entities and the number of available codes.

Figure 1(b) shows a schematic diagram of a multi-bit magnetic tag. The tag consists of several thin film magnetic elements supported by a polymer substrate. The magnetic elements are elongated perpendicular to the long axis of the substrate, and therefore exhibit uniaxial shape anisotropy, so that their magnetisation is only stable if it lies dominantly along the long axis. Each element can therefore support two oppositely aligned magnetisation states which form the '1's and '0's of a binary code written along the tag.

As explained previously, for a SAT system to offer sufficiently high throughput it is desirable that the tags can be read while in microfluidic flow. The most attractive method of achieving this for multi-bit magnetic tags is to sequentially measure the orientation of the magnetisation in each of the magnetic elements comprising the tag using a highly sensitive magnetic field sensor embedded beneath the floor of the microfluidic channel through which the tags are flowing.

To assess the feasibility of reading the tags in this fashion a micromagnetic model was created to investigate how a magnetic sensor would respond to the stray field patterns generated by the tags. Shen *et al.* have previously used a tunnel magnetoresistance (TMR) sensor to detect the presence of individual micron-sized magnetic beads with excellent signal-to-noise ratios [23]. The model of the sensor was therefore designed to replicate the behaviour of this device.

It was found that the magnetoresistance curve of the sensor could be reproduced by using a simple Stoner-Wohlfarth model to represent the switching behaviour of the sensor's free layer. Figure 5(a) compares the experimentally measured transfer curve of the sensor, taken from [23], with that calculated using the model. It can be seen that the experimental and theoretical data are in good agreement.

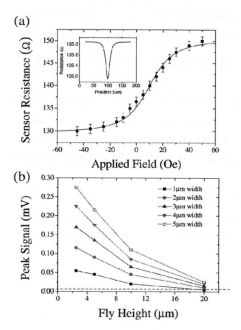

FIGURE 5. (a) Simulated transfer curve of the TMR sensor (solid line). The data points are experimental results taken from the work of Shen *et al.* [23]. The inset figure shows the response of the sensor to a rectangular NiFe element with a length of 20 μm, a width of 3 μm and a thickness of 20 nm. The element is scanned over the sensor at a height of 2.5 μm with its long axis parallel to the sensor's easy axis. (b) Peak sensor signal against fly height for 20 μm long, 20 nm thick NiFe elements of various widths. The dashed red line indicates the sensor's noise floor as quoted in [23].

The magnetic field patterns produced by the tags were calculated using finite-element simulations. The simulated elements were 20 μm long and composed of a 20 nm thick NiFe film which was assumed to be uniformly magnetised. Elements with widths in the range 1 to 5 μm were considered, and were initially studied in isolation to enable the detection limits of the system to be evaluated. The field pattern produced by each of these elements was passed over the simulated sensor at a variety of fly heights, with the long axes of the elements aligned parallel to the sensor's easy axis. An example TMR response is shown in the inset to Figure 5(a). A clear dip is visible as the tag passes over the position of the sensor, which is situated 100 μm from the tag's initial position.

Figure 5(b) shows how the peak signal produced by each of the NiFe elements varies with their fly height above the sensor. The dashed red line indicates the 5 μV noise floor which is quoted for the sensor device. Signals in mV were calculated by assuming that the TMR sensor was placed within a balanced AC bridge as in [23]. The shape of the MR peaks were found to be well described by Lorentzian distributions, the widths of which increased with the fly height of the elements, but showed only weak dependence on the width of the elements.

When the fly height of the tag was 20 μm, the 5 μm wide element is detected with a signal-to-noise ratio of 5:1, however the signal from the 1 μm wide element was only 5 mV, equal to the noise floor of the device. When the fly height of the tag is reduced to 10 μm all of the elements can be detected with signal-to-noise ratios greater than 4:1. This demonstrates that using a magnetoresistive sensor to read the codes from the magnetic barcode tags should be possible, provided that the fly height of the tags can be adequately controlled.

In order to determine how many unique codes planar magnetic tags could accommodate it is necessary to know how widely the individual bits of the tag must be spaced, if they are to be individually resolved by the sensor. To investigate this, simulations were performed where a five bit tag was passed over the sensor. A schematic diagram of the simulated tag is shown in Figure 6(a). The elements in the tag had widths of 1, 2, 3, 4 and 5 μm and other dimensions as in the previous simulations. The aspect ratios of the elements comprising the tag are varied in this way so that each element has a different coercivity and will switch at a different applied field, allowing the entire tag to be written using a non-local field of varying amplitude, as it is demonstrated in a separate publication [24].

Figures 6(b)-6(d) show how the simulated sensor responded when the five bit tag was passed over it at a variety of fly heights. The spacing between the elements was set to be 15 μm. Data for tags carrying the (11111) and (10101) codes is shown. While the (10101) code is resolved well at all fly heights, it is clear that for the (11111) code the individual peaks become progressively harder to resolve as the fly height of the tags is increased. This is due to the widths of the MR peaks from each of the elements in the tag becoming larger at greater fly heights, as was previously discussed.

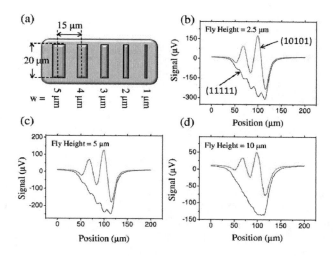

FIGURE 6. Simulated sensor output in response to a five-bit planar magnetic tag. (a) Schematic diagram showing the geometry of the tag. Each element is composed of NiFe and is 20 nm thick. (b)-(d) Simulated sensor response when the tag is passing (b) 2.5 μm, (c) 5 μm and (d) 10 μm above the sensor. Data for tags carrying both the (11111) and (10101) codes are shown.

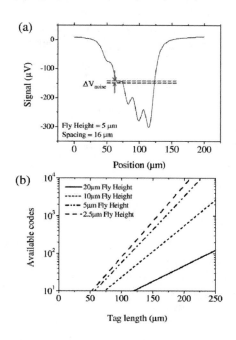

FIGURE 7. (a) Simulated sensor response from a five-bit magnetic tag in which the inter-element spacing is set to be equal to the FWHM of the single element response at a fly height of 5 μm. The sensors noise level, as quoted in [23], is also shown. (b) Graph showing code carrying capacity as a function of tag length for maximum fly heights of 20 μm, 10 μm, 5 μm and 2.5 μm.

121

Figure 7(a) shows the signal produced by a five bit tag passing above the sensor at a fly height of 5 μm. The tag was in the (11111) configuration. The separation between the elements in the tag was set to be 16 μm, which is approximately equal to the Full Width at Half Maximum (FWHM) of the MR peaks produced by the elements at a 5 μm fly height. It can be seen that the peaks corresponding to each of the magnetic elements can now be resolved. This suggests that separating the magnetic elements by the FWHM of their MR peaks is a suitable criterion for calculating the code-carrying capacity of the multi-bit tags. However, it should also be noted that in a real tag it might be beneficial to use non-uniform element spacing so that the signals from the smaller magnetic elements do not become obscured by their larger counterparts, as can be seen to occur to some degree in Figure 7(a).

Figure 7(b) shows how the calculated code-carrying capacity of the planar magnetic tags varies with the substrate length for a variety of fly heights. With a maximum fly height of 20 μm a substrate with a length of approximately 200 μm can accommodate 5 bits and therefore 32 different codes. The 200 μm substrate we assume here is similar to the particle size in other high-throughput SAT schemes [10]. If the fly height can be regulated to be no more than 10 μm the same substrate can accommodate 9 bits and therefore 512 codes. At a 5 μm fly height, 12 bits can be accommodated and therefore the tags can carry 4096 different codes. Finally, if the fly height can be controlled such that no tag has a fly height greater than 2.5 μm, 13 bits can be placed on the substrate allowing for 8192 distinct codes. Planar magnetic barcode tags therefore offer a much higher degree of assay parallelism than could be achieved using a simple moment encoded approach to tagging.

CONCLUSIONS

In this paper we have used micromagnetic simulations to examine the feasibility of creating a high-throughput suspension assay technology, in which the identities of a biocompatible particle's surface receptors are encoded using magnetic signatures.

In the first approach the tags are superparamagnetic microbeads similar to those which are commercially available. The beads are encoded by varying their magnetic content, and therefore their magnetic saturation moment.

Micromagnetic simulations have been presented which indicate that it should be possible to determine the moment of the magnetic beads as they flow past a digital magnetic sensor. However, when we introduce a realistic distribution of bead moments into each bead subset and allow for some uncertainty in the path of the beads over the sensor, it is found that the sensor's ability to unambiguously identify each bead subset is diminished. This limits the number of unique tags that can be made available, and hence the number of biological assays that can be multiplexed. In our case study we find that, allowing for a 5% error rate, only four subsets of beads could be differentiated. This demonstrates the inefficiency inherent to a tagging system based on the total magnetic moment of the tags. However, assays where only a small number of unique probes are required could be performed using such a technology.

In the second approach the tags are multi-bit planar magnetic tags, consisting of a number of elongated thin film magnetic elements supported by a polymer substrate. A binary code is written along the substrate using the two stable magnetisation states of the magnetic elements.

We have performed micromagnetic simulations showing that the stray fields produced by the magnetic elements are detectable using a contemporary TMR sensor device. The code carrying capacity of the tags has also been investigated, and our studies indicate that a 200 μm long substrate could carry as many as 13 distinguishable magnetic bits and therefore provide over 8000 unique codes. This exceeds the number of codes that can be generated using a simple fluorescent tagging system by over an order of magnitude [3], and is competitive with other SAT technologies which use more complex optical tagging techniques, such as quantum dots [5, 25]. It is also likely that bit density could be increased beyond what is suggested here by careful optimisation of the spacing between each pair of neighbouring bits.

SAT technology based on magnetic planar tags has advantages over forms of SAT which use optical encoding. Many biological samples exhibit a high optical background (for instance due to autofluorescence) from which the tags' signatures must be extracted. In contrast to this, biological samples exhibit an extremely low magnetic background. Most optical SAT technologies also require image analysis to interpret optical microscopy images of the tags. Potentially, this is computationally expensive and could limit a systems serial throughput. The magnetic tags we propose are read using a conventional TMR sensor similar to those used as read heads in contemporary hard disks, and therefore we anticipate that little processing will be required to reconstruct the tag's code from the sensor's output. Furthermore, TMR sensors are much cheaper to produce than high speed CCD camera equipment. Our magnetic SAT technology should therefore be cheaper and capable of a higher throughput than optical SAT technology.

Further advantages over existing SAT technology are presented by the fact that the planar magnetic barcodes are writable. In principle all of the tags are structurally identical, differing only by nature of their surface receptors. This will make them much cheaper to produce than tags which require a slightly different fabrication process for each distinct tag subset [2-9]. Assay flexibility is also improved by making the tags writable because new probe sets can be assigned to any code that is currently unused in the assay.

A further exciting possibility is that writable tags with high numbers of magnetic bits could be combined with "split and mix" combinatorial chemistry techniques to build up huge libraries of probe biomolecules, such as DNA [26, 27]. High throughput assays could then be performed with these libraries, allowing DNA to be sequenced at previously unimaginable rates [28]. Such a technology could potentially usher in a new era of personal genetic sequencing.

ACKNOWLEDGEMENTS

The authors acknowledge the Engineering and Physical Sciences Research Council (EPSRC) for financial support. We also thank G. Xiao and J. Fonnum for useful advice and discussions.

REFERENCES

1. J.P. Nolan and L.A. Sklar, *Trends. Biotechnol.* **20** (1), 9-12 (2002).
2. T.M. McHugh, R.C. Miner, L.H. Logan and D.P. Stites, *J. Clin. Microbol.* **26** (10), 1957 (1988).
3. R.J. Fulton, R.L. McDade, P.L. Smith, L.J. Kienker and J.R. Kettman, *Clin. Chem.* **43** (9), 1749 (1997).
4. E.J. Moran, S. Sarshar, J.F. Cargill, M.M. Shahbaz, A. Lio, A.M.M. Mjalli and R.W. Armstrong, *J. Am. Chem. Soc.* **117**, 10787 (1995).
5. S.R. Nicewarner-Peña, R.G. Freeman, B.D. Reiss, L. He, D.J. Peña, I.D. Walton, R. Cromer, C.D. Keating and M.J. Natan, *Science* **294**, 137 (2001).
6. M. Han, X. Gao, J.Z. Su and S. Nie, *Nature Biotechnol.* **19**, 631 (2001).
7. F. Cunin, T.A. Schmedake, J.R. Link, Y.Y. Li, J. Koh, S.N. Bhatia and M.J. Sailor, *Nature Mat.* **1**, 39 (2002).
8. K. Braeckmans, S.C. De Smedt, C. Roelant, M. Leblans, R. Pauwels and J. Demeester, *Nature Mat.* **2**, 169 (2003).
9. G.S. Galitonov, S.W. Birtwell and N.I. Zheludev, *Optics Express* **14** (4), 1382 (2006).
10. D.C. Pregibon, M. Toner and P.S. Doyle, *Science* **315**, 1393 (2007).
11. D.R. Baselt, G.U. Lee, M. Natesan, S.W. Metzger, P.E. Sheehan and R.J. Colton, *Biosensors Bioelectr.* **13**, 739 (1998).
12. M. Megans and M.W.J. Prins, *J. Magn. Magn. Mat.* **293**, 702 (2005).
13. M.M. Miller, P.E. Sheehan, R.L. Edelstein, C.R. Tamanaha, L. Zhong, S. Bounak, L.J. Whitman and R.J. Colton, *J. Magn. Magn. Mat.* **225**, 138 (2001).
14. H.A. Ferreira, D.L. Graham, N. Feliciano, L.A. Clarke, M.D. Amaral and P.P. Freitas, *IEEE Trans. Magn.* **41** (10), 4140 (2005).
15. J. Llandro, T.J. Hayward, D. Morecroft, J.A.C. Bland, F.J. Castaño, I.A. Colin and C.A. Ross, *Appl. Phys. Lett.* **91**, 203904 (2007).
16. T.J. Hayward, J. Llandro, R.B. Balsod, J.A.C. Bland, D. Morecroft, F.J. Castaño and C.A. Ross, *Phys. Rev. B.* **74**, 134405 (2006).
17. F.J. Castaño, D. Morecroft, W. Jung and C.A. Ross, *Phys. Rev. Lett.* **95**, 137201 (2005).
18. T.J. Hayward, J. Llandro, F.D.O. Schackert, D. Morecroft, R.B. Balsod, J.A.C. Bland, F.J. Castaño and C.A. Ross, *J. Phys. D.: Appl. Phys.* **40**, 1273 (2007).
19. OOMMF software package, M. Donahue and D. Porter, http://math.nist.gov/oommf/.
20. G. Fonnum, C. Johansson, A. Molteberg, S. Mørup and E. Aksnes, *J. Magn. Magn. Mat.* **293** (1), 41 (2005).
21. G. Fonnum (private communication).
22. K. van Ommering, J.H. Nieuwenhuis, L.J. van IJzendoom, B. Koopmans and M.J.W. Prins, *Appl. Phys. Lett.* **89**, 142511 (2006).
23. W. Shen, X. Liu, D. Mazumdar and G. Xiao, *Appl. Phys. Lett.* **86**, 253901 (2005).
24. J.R. Jeong, J. Llandro, B. Hong, T.J. Hayward, T. Mitrelias, K.P. Kopper, T. Trypiniotis, S.J. Steinmuller, G.K. Simpson and J.A.C. Bland, *submitted to Lab-on-a-chip*.
25. P.S. Eastman, W. Ruan, M. Doctolero, R. Nuttall, G. de Feo, J.S. Park, J.S.F. Chu, P. Cooke, J.W. Gray, S. Li and F.F. Chen, *Nano Letters* **6**, 1059 (2006).
26. A.R. Vaino and K.D. Janda, *PNAS* **97** (14), 7692 (2000).
27. B.J. Battersby, D. Bryant, W. Meutermans, D. Matthews, M.L. Smythe and M. Trau, *J. Am. Chem. Soc.* **122**, 2138 (2000).
28. T. Mitrelias, T. Trypiniotis, F. van Belle, K.P. Kopper, S.J. Steinmuller, J.A.C. Bland and P.A. Robertson, same volume.

Detection of Magnetic-Based Bio-Molecules Using MR Sensors

Marius Volmer[a] and Marioara Avram[b]

[a]Physics Department, Transilvania University, 29 Eroilor, Brasov 500036, Romania
[b]National Institute for Research and Development in Microtechnologies, Str. Erou Iancu Nicolae 32B,
72996 Bucharest, Romania

Abstract. Results from micromagnetic modelling performed on magnetoresistive biosensors are presented in this paper. The biosensors consist of giant magnetoresistive (GMR) type multilayers. The magnetic beads labelled with different biological structures are assembled on top of the sensor element or between sensor elements. In both cases we used differential measurement setup in order to extract only the signal which corresponds to the field produced by the magnetic beads. The simulations were performed when the magnetic field is applied parallel and perpendicular to the easy axis of the GMR sensor. When the magnetic particles are located between two GMR sensors, placed in opposite branches of a Wheatstone bridge, sharp and symetric peaks will appear around zero field at which the GMR sensors present a maximum sensitivity.

Keywords: Magnetoresistive sensors, Magnetic beads, Magnetic microsystems.
PACS: 75.47.De, 75.70.Ak, 87.85.fk, 87.85.Ox

INTRODUCTION

Lab-on-a-chip devices [1-3] shrink entire chemical or biochemical assays down to small microfluidic chips. Because MR sensors made with magnetic layers are sensitive for small magnetic fields, they are very attractive for lab-on-a chip applications used for biological diagnostics. This application includes detection of small magnetic fields generated by magnetic particles encapsulated in plastic, carbon or ceramic spheres which are coated with chemical or biological species such as DNA or antibodies that selectively bind to the target analyte. Results from theoretical modelling, as well as laboratory tests, show that GMR detectors can resolve single micrometer-sized magnetic beads [4-6]. Figure 1 shows [4, 5], in a simple view, the bonding of the beads to the sites - GMR sensor - via the molecules to be detected (antigens, in our example). Several bioassays can be simultaneously accomplished using an array of magnetic sensors, each with a substance that bonds to a different biological molecule [5, 7-9]. This application requires extremely small, low-power, low field magnetic sensors. The magnetic microbeads and the GMR sensors will be coated with identical probes such as DNA or antibodies which are specifically bonded together by the correct analyte molecule to be analyzed. The microbeads in suspension were allowed to settle onto the GMR sensor array where specific beads bonded to specific sensors

CP1025, *Biomagnetism and Magnetic Biosystems Based on Molecular Recognition Processes*
edited by J. A. C. Bland and A. Ionescu
© 2008 American Institute of Physics 978-0-7354-0547-9/08/$23.00

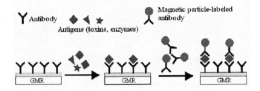

FIGURE 1. Antigens are detected by flowing them over a sensor coated with antibodies to which they bind. The magnetic particle-labeled antibodies then bind to the antigens providing a magnetic indication of the presence of those antigens.

only if the probes were designed to attract each other. Nonbinding beads can be removed by applying a small magnetic field. The beads are then magnetized by a DC or an AC electromagnet and detected by the GMR sensors. The microbeads are made up of nm sized iron oxide particles that have little or no magnetization in the absence of an applied field.

We used a micromagnetic simulator, SimulMag [10], to obtain the response of the magnetoresistive sensors

RESULTS AND DISCUSSION

In this section we describe the detection of magnetic beads labeled with different biological structures. Magnetic micro-beads when magnetized by an external field have a magnetic dipole field as sketched in Figure 2a. This field is proportional to the volume of magnetic material and inversely proportional to the distance cubed from the bead to the sensor which makes difficult the detection process. On the other hand, if the sensitive area is much larger than the bead, only the portion of the magnetoresistive material close to the bead will be affected. Therefore the fractional change in resistance, and hence the sensitivity, will be maximized by matching, as far as possible, the size of the sensor to the size of the bead and by reducing the distance between the sensor and the bead. It was shown that single magnetic markers of any size can be detected as long as the GMR sensor has about the same size as the marker and the insulating protection layer is thin enough [5].

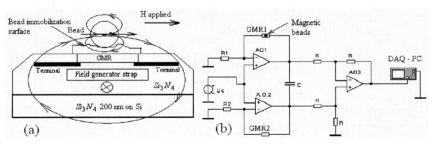

FIGURE 2. (a) The schematic setup for a GMR sensor that detects the stray field generated by a bead. The field that magnetizes the immobilized bead can be generated by a strap located beneath the sensor. (b) The differential measurement setup used to measure the stray field produced by the magnetic bead. The sensor GMR2 is used as reference.

The amplitude of stray fields produced by the beads depends on the size and magnetic moment of the particles.

In our simulations regarding the sensor response we'll assume a total thickness of the layers (immobilization layer and protection layer – Si_3N_4) between the bead and the GMR sensor of 0.2 μm. The sensor dimensions are $1x1x0.1\mu m^3$. The bead is assumed to be a sphere with a diameter of about 0.5 μm. Because the magnetization of magnetite is lower by a factor of 2–3 than for other ferromagnetic materials (e.g. cobalt or iron) [5], we assumed, in our simulations, for the saturation magnetization of the micro-magnetic bead a value of 400 emu/cm^3 (the saturation magnetization for Co is 1300 emu/cm^3). For an external magnetic field of 500 Oe, the amplitude of the stray field in the sensor region, under the above conditions, is between 8 and 24 Oe if the size of the bead is ranging from 0.35 μm to 0.6 μm [5]. In order to extract the contribution corresponding to the total magnetic moment of the bead and to avoid the influence of the external magnetic fields and thermal variations we designed a differential measurement system as shown in Figure 2b. The driving voltage, U_C, sets constant currents, $I_1=U_C/R_1$, $I_2=U_C/R_2$, through the sensors GMR1 and GMR2. From basic electronics, we have for the output voltage that is applied to the data acquisition system the expression (in modulus):

$$\Delta U = \left(U_{GMR1} + U_C\right) - \left(U_{GMR2} + U_C\right) = U_{GMR1} - U_{GMR2} = f\left(H_{bead}\right), \qquad (1)$$

where H_{bead} denotes the strength of the stray field generated by the magnetic bead in the sensor region.

The results of our micromagnetic simulations for a differential measurement system, according to physical and electronic setup shown in Figure 2, are presented in Figure 3a when the magnetic field is applied parallel to the easy axis of magnetization and perpendicular to the easy axis in Figure 3b respectively. The results show only the contribution of the stray fields produced by the micro-bead. The output, ΔU, of the differential amplifier is directly proportional to the changes of the GMR amplitude that takes place in the sensor GMR1 due to the presence of the micro-bead at 200 nm above.

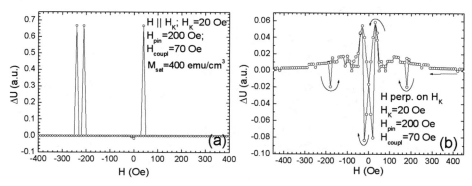

FIGURE 3. The response of a differential system with GMR sensors for a magnetic micro-bead placed 200 nm above the sensor GMR1, when the magnetic field is applied (a) parallel and (b) perpendicular to the easy axis. The sensor GMR2 is used as reference. The arrows are guide for the eyes.

Because the output voltage, ΔU, depends on some parameters like the amplitude of the GMR effect, the resistance value of the GMR sensors, the driving currents through the sensors and the gain of the instrumentation amplifier, the vertical axis is quoted in arbitrary units (a.u.). In these simulations the GMR sensors consist of multilayer structures as FeMn/NiFe(2 nm)/Cu(1 nm)/NiFe(2 nm) for which we assumed a small positive coupling between the magnetic layers through the nonmagnetic layer, H_{coupl}=70 Oe and a uniaxial anisotropy field H_K=20 Oe. This positive coupling is often present in real structures due to the small iregularities of the surfaces giving the so called orange-peel coupling [11]. In order to obtain the GMR effect in these structures, the magnetization of bottom layer is pinned by exchange interaction, using a layer of FeMn. In this simulation, the pinninig field was set at 200 Oe, a value which is consistent with experimental data. The bead is placed above the free layer. The amplitude of the $\Delta U(H)$ plots and positions of the peaks depend on the volume and magnetic moment of the micro-beads.

A few comments are necessary regarding the differences between the curves presented in Figures 3a and 3b. When the magnetic field is applied upon the easy axis, the magnetization from the free layer will reverse at a value which is dependent on the anisotropy field, the coupling field and the strength of the magnetostatic interaction between the free layer and the magnetic bead. The $M(H)$ and $GMR(H)$ characteristics present an asymmetry around zero field which is typical for exchange biased spin valves. The presence of the magnetic beads above the sensor GMR1 will bring an additional field, H_{bead}, which alters the $GMR(H)$ response. These perturbations can be seen using a reference sensor, GMR2, in a differential setup as we presented in Figure 2. The positions of these peaks are mainly related with the anisotropy and pinning fields. Our simulations revealed the fact that the position of the low field peak, depends also on the surface coverage ratio with magnetic beads. For this reason the best detection method is to sweep the magnetic field between the limits illustrated in Figure 3, i.e. to perform a complete magnetization curve. When the magnetic field is applied perpendicular to the easy axis, the $M(H)$ and $GMR(H)$ characteristics are symmetrical around zero field. Consequently, the differential response which gives the contribution of magnetic beads will be a symmetrical one as was obtained by micromagnetic simulations and presented in Figure 3b. The peaks are located around the anisotropy field (20 Oe) and pinning filed (200 Oe).

Using alternating magnetic fields it is possible to compensate the thermal drift of the sensors and electronics and the structure temperature remains at low values avoiding the damaging of the biological molecules.

Another method to measure the magnetic field produced by the micro-beads is to use a full Wheatstone bridge in which all the resistors are GMR elements subjected to the same external magnetic field, H. The magnetic particles will be located between two sensors, GMR1 and GMR2, placed in opposite arms of the bridge, as in Figure 4. The sensors GMR3 and GMR4 are used as reference sensors. Using this setup, the output voltage, ΔU, will be a function of the magnetic field produced by the beads.

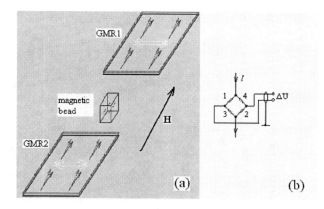

FIGURE 4. The measurement setup (a) and the electrical connection of the sensors (b). GMR1 and GMR2 are active sensors whereas GMR2 and GMR4 are used as reference sensors.

Using the same data for sensors and magnetic bead the micromagnetic simulations give us the response of the measurement circuit, presented in Figure 5.

The measurement configuration presented above has the advantage of sharp and symetric peaks around zero field at which the GMR sensors presents a maximum sensitivity. The particle detection will take place in low magnetic fields which means a smaller electrical current through the field generator strap and a lower temperature of the chip.

CONCLUSIONS

We presented in this paper some micromagnetic simulations regarding the use of the GMR sensors for biomedical applications such as detection of magnetic particles labelled with biological systems that have to be investigated. The results of these simulations will be verified by experiments using measurement systems that we designed and presented in this paper.

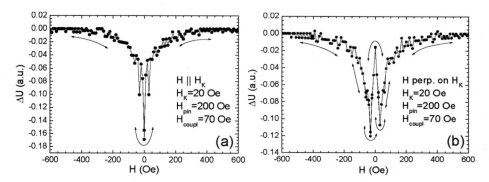

FIGURE 5. The response of the system, presented in Fig. 4, when the easy axis of the GMR sensors is (a) parallel and (b) perpendicular to the applied magnetic field. The arrows are guides for the eyes.

ACKNOWLEDGMENTS

This work was supported by grant CEEX 27/10-10-2005.

REFERENCES

1. D. Figeys and D. Pinto, *Anal. Chem.* **72** (9), 330A–335A (2000).
2. M.A.M. Gijs, *Microfluidics and Nanofluidics* **1**, 22–40 (2004).
3. C. Ahn and J.-W. Choi, "Microfluidics and Their Applications to Lab-on-a-Chip", in *Springer Handbook of Nanotechnology*, edited by Bharat Bhushan, Berlin: Springer Verlag, 2004, pp 270-277.
4. M. Tondra, M. Porter and R.J. Lipert, *Journal of Vacuum Science and Technology A: Vacuum, Surfaces, and Films* **18** (4), 1125-1129 (2000).
5. J. Schotter, P.B. Kamp, A. Becker, A. Pühler, G. Reiss and H. Brückl, *Biosensors and Bioelectronics* **19**, 1149-1156 (2004).
6. M. Brzeska, M. Panhorst, P.B. Kamp, J. Schotter, G. Reiss, A. Pühler, A. Becker and H. Brückl, *Journal of Biotechnology* **112** (1-2), 25-33 (2004).
7. L.J. Kricka, *Clin. Chim. Acta* **307**, 219–223 (2001).
8. M.M. Miller, P.E. Sheehan, R.L. Edelstein, C.R. Tamanaha, L. Zhong, S. Bounnak, L.J. Whitman and R.J. Colton, *J. Magn. Magn. Mater.* **225**, 138–144 (2001).
9. H. A. Ferreira, D. L. Graham and P. P. Freitas, *J. Appl. Phys.* **93** (10), 7281 (2003).
10. J.O. Oti, "SimulMag Version 1.0" in *Micromagnetic Simulation Software, User's Manual*, Boulder, Colorado: Electromagnetic Technology Division, National Institute of Standards and Technology, 1997.
11. J.C.S. Kools, T.G.S.M. Rijks, A.E.M. De Veirman and R. Coehoorn, *IEEE Trans. Magn.* **31**, 3918 (1995).

Giant Magnetoimpedance for Biosensing in Drug Delivery

Vanesa Fal-Miyar[a], Arun Kumar[b], Shyam Mohapatra[b], Shawna Shirley[b], Natalie. A. Frey[c], José M. Barandiarán[d], and Galina V. Kurlyandskaya[d]

[a]Department of Physics, University of Oviedo, Avd/ Calvo Sotelo s/n, 33007, Oviedo, Spain
[b]Department of Internal Medicine, Division of Allergy and Clinical Immunology, University of South Florida, Tampa, FL 33612, USA
[c]Department of Physics, PHY 114, 4202 East Fowler Ave, University of South Florida, Tampa, FL33620
[d]Department of Electricity and Electronics, University of Basque Country UPV-EHU, Apartado 644, 48080 Bilbao, Spain

Abstract. Iron oxide (Fe_3O_4) non-specific superparamagnetic nanoparticles of 30 nm size are introduced into human embryonic kidney (HEK-293) cells by intracellular uptake. The nanoparticles are magnetised by two superimposed magnetic fields, an externally applied DC field and an AC field generated by the high-frequency current flowing through $Co_{64.5}Fe_{2.5}Cr_3Si_{15}B_{15}$ amorphous ribbons. The resulted fringe fields from the nanoparticles are detected via the magnetoimpedance change in the ribbons covered and uncovered by thin gold layer. The gold covering is considered an improvement due to its biocompatibility and because it avoids the biocorrosion process on the ribbon. The MI responses in both cases are clearly dependent on the presence of the magnetic nanoparticles inside the cells and on the value of the external field.

Keywords: Magnetoimpedance, Biosensor, Nanoparticles, Intracellular up-take.

PACS: 75.47.-m, 87.85.fk, 78.67.Bf

INTRODUCTION

In recent years, biosensors based on magnetic effects are becoming common tools in biomedicine and environmental control [1-6]. In such biosensors, the associated or integrated transducers are made from magnetic materials. Different magnetic effects can be used for design of a magnetic biodetector. The magnetoimpedance (MI) has been proposed as one of the techniques which have a capacity to sense the presence of magnetic particles [3] and has been used to detect biomolecular labels in different kinds of biological fluids [3, 6].

The magnetoimpedance is based on the change of the high frequency impedance of a ferromagnetic conductor under the application of a magnetic field [7]. In the particular case of the MI biosensor prototype, based on magnetic label detection, the change of the impedance response of the sensitive element immersed in a buffer solution which contains magnetic labels was analyzed under application of an external

CP1025, Biomagnetism and Magnetic Biosystems Based on Molecular Recognition Processes
edited by J. A. C. Bland and A. Ionescu
© 2008 American Institute of Physics 978-0-7354-0547-9/08/$23.00

magnetic field [6]. The functionality basis is similar to that proposed earlier for biosensors working on the principle of giant magnetoresistance (GMR) [1]: a fringe magnetic field induced by the magnetic particles works as a physical transducer. In all reported GMR biosensors the magnetic labels are fixed at a specific distance from the magnetic sensitive element [1, 8], the feature which allowed simple data evaluation. In 2005 the idea of a sensing process with non-fixed magnetic labels was reported and MI was proposed for evaluation of the quantity of magnetic nanoparticles embedded into living cells by up-take events. This suggestion was based on the fact that MI field sensors shows very high sensitivity to a magnetic field in comparison with other magnetic detectors (up to 450 %/Oe) [6]. Since then no experimental study has been done to develop this particular type of biodetector.

In this work, we have designed the MI biosensor prototypes for the detection of the presence of superparamagnetic nanoparticles introduced into cells by intracellular uptake. The MI responses of Co-based amorphous ribbon as sensitive element non-covered or covered by thin gold layer were studied for a model system of embryonic kidney (HEK 293) cells and Fe_3O_4 non specific magnetic nanoparticles.

EXPERIMENTAL SETUP

For the sensitive element a single-roller quenched $Co_{64.5}Fe_{2.5}Cr_3Si_{15}B_{15}$ amorphous ribbons, with dimensions 0.02 x 1.2 x 10 mm^3 were selected. This material shows very high MI effect for relatively low driving current intensity and very high stability to a biocorrosion [5]. Afterwards two sets of sensitive elements were used: as quenched CoFeCrSiB amorphous ribbons and as quenched CoFeCrSiB ribbons covered by 150 nm gold layers deposited by rf-sputtering onto both surfaces. The longitudinal hysteresis loops of the as quenched and gold covered ribbons were carried out by an inductive method for 10 Hz magnetic field cycling.

The MI was measured by a four probe technique with constant amplitude of the sinusoidal current in the range of 5 to 50 mA. The experimental system is shown in Fig. 1. Two power supplies (Fig. 1(a)) - Keithley 2400 and Kepko ATE-DMG) were used to feed a pair of Helmholtz coils to create a magnetic field in the plane of the ribbon. The selected driving current frequency from the range of 0.5 to 10 MHz was kept constant for each particular MI curve. To adjust the current amplitude, a non-inductive resistor was connected in series with the ribbon (Fig. 1 (inset)). At each field point, the intensity of the driving current created by an Agilent 33220A (Fig. 1 (b)) function generator was measured and the amplitude was tuned to a desirable value. A Tektronix TDS3032 (Fig. 1 (c)) oscilloscope was used to measure the voltage drop across the sample. The MI ratio ($\Delta Z/Z$) was defined as follows:

$$\frac{\Delta Z}{Z} = 100 \times \frac{[Z(H) - Z(H_{max})]}{Z(H_{max})}, \tag{1}$$

where the maximum values of the applied magnetic field were $H_{max} = \pm 78$ Oe. The frequency of 5 MHz was selected for studies with biological samples because the corresponding MI ratio was close to its maximum value in the 0.5 - 10 MHz range.

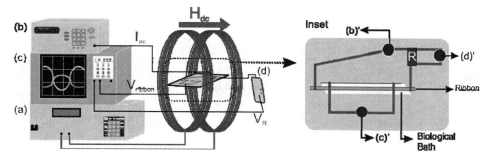

FIGURE 1. Schematic representation of MI setup: (a)- Keitley and Kepko power supplies to feed the Helmholtz coils, (b)- Agilent 33220 A function generator to apply the I_{ac} current connected at (a)' by a BNC, (c)- Tektronix TDS3032 oscilloscope to measurement the difference of potential connected at (b)' and (d)- differential probe to measure the differential signal of voltage drop across the reference resistor (d)'. Inset: A magnification of the sample holder.

The testing amplitudes of the current were selected in the range of 5 to 50 mA to get a sufficient MI sensitivity and to avoid undesirable heating of the biological samples. The sensitive element was installed in the central part of the imprinted circuit using an Ag conductive paint (Fig. 1(inset)).

For the measurements in presence of the biological samples, a non-magnetic bath of 1.5 ml was situated in the centre so that both surfaces of the ribbon were in contact with the HEK 293 cells. The HEK 293 cells were generated by transformation of human embryonic kidney cell cultures (hence HEK) with sheared adenovirus 5 DNA [9]. These cells were chosen because they are a well known and widely used cell line in cell biology research showing high level of non-specific up-take. The HEK-293 cells were cultivated to 80% of confluence (growth of cells in a monolayer plane until they come in contact with each other) in Dulbecco's Modified Eagle's Minimal Essential Medium, DMEM, (with 4.5 g/l of L-glutamine and sodium pyruvate) supplemented with 10% fetal bovine serum (FBS) and 1% Penicillin-Streptomycin solution (P/S). The culture medium was replaced every 48 h with DMEM supplemented with FBS and P/S [10].

Then the HEK 293 cells were infected with Fe_3O_4 magnetic nanoparticles. Approximately one million cells in 2 ml of culture media were seeded into culture dishes and allowed to reach 80% confluency. The cells used for this experiment were passed through 19 generations and the pH of the culture was 7.9. To infect the cells, the Fe_3O_4 magnetic particles were prepared by method reported by Mandal *et al.* [11]. The iron solution was made using ferrous ammonium sulphate and ferric ammonium sulphate with the molar ratio 1:2 and 0.4M of aqueous sulphuric acid. Meanwhile 1M NaOH and 1M TritonX-100 were prepared to be mixed. This solution was heated at 80°C and the shock solution was put in it drop wise. When the solution started to become brown it was centrifuged and the particles were taken and dried in a vacuum oven. Afterwards, 100 ml of particle suspension at a concentration of 0.05 mg/ml of particles in phosphate buffered saline (PBS) and a pH of 7.2 was introduced into the cell culture. The nanoparticles were characterized by XRD and the phase purity was verified by comparing the data with standard Fe_3O_4. The size of the Fe_3O_4 particles

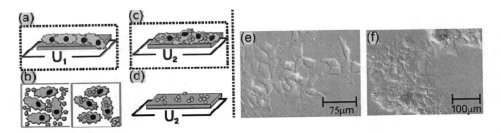

FIGURE 2. (a-d) Schematic representation of testing: (a)- The voltage drop across the MI sensitive element is measured in presence of solution of cells (b)- The infection where the cell were mixed with the magnetic nanoparticles, which was introduced into the cells by uptake events and the supernatant, free nanoparticles, were removed. (c)- The voltage drop across the MI element installed in a bath which contains the magnetic nanoparticles inside the cell. (d)- The magnetic equivalent of 3 (e, f). Optical image taken before MI measurement of: (e)- HEK 293 cell (f)- HEK 293 cells infected by nanoparticles.

was estimated using a Transmission Electron Microscopy (TEM). The average size of the particles was about 30 nm. The magnetic characterization of the magnetic particles inside the cells was performed by a Quantum Design Physical Properties Measurement System (SQUID) at room temperature. Optical microscopy and TEM were used for temperature. Optical microscopy and TEM were used for characterization of the biological samples before the MI measurements.

RESULTS AND DISCUSSION

Figure 2 (a-d) shows the scheme of the testing procedure. As a first step, MI of the sensitive element installed in a bath for fluids is measured in the presence of a certain number of cells placed in a culture medium (Fig. 2 (a)). The corresponding experimental parameter is a voltage drop U_1 from which an impedance value can be calculated using Ohm's law. This first step is actually a calibration of the sensor for zero concentration of the nanoparticles. The second step is the biochemical part of testing (Fig. 2 (b)). Magnetic nanoparticles are mixed with the cell sample. After incubation, a certain number of nanoparticles get introduced into cells by uptake events and the remaining free magnetic nanoparticles are removed. As the third step, MI of the sensitive element installed in a bath for fluids is measured in the presence of the nanoparticle-loaded cells (Fig. 2 (c)). The corresponding experimental parameter is a voltage drop U_2. In a simple linear response model the difference of the induced voltages U_1 and U_2 represents the quantity of nanoparticles participating in the uptake events. A detailed analysis of the sensor response requires an elaborate mathematical modelling for concentration definition because in this regime not only the fringe field of each particle but also their spatial distribution. Figure 2 (d) shows the physical equivalent (magnetic nanoparticles in a particular spatial configuration) of the biological testing represented in Fig. 2 (c). The optical images of the cells before and after transfection can be seen in Fig. 2 (e, f), respectively. Detection in the described

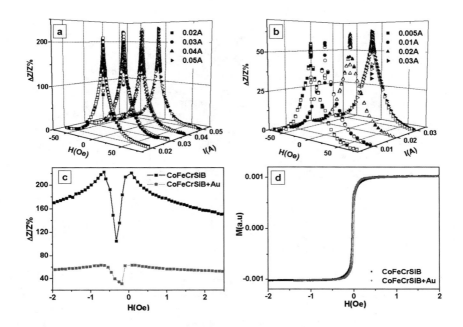

FIGURE 3. (a) MI responses of CoFeCrSiB for different alternating current intensities and a frequency of 5 MHz. (b) MI responses of the CoFeCrSiB ribbon covered by gold for different alternating current intensities and a frequency of 5 MHz. (c) MI response in decreasing field for alternating current intensity of 30mA for the amorphous ribbon and the composite. (d) Longitudinal hysteresis loops of the amorphous ribbon and the composite carried out by the inductive method.

system allowed to define the quantity of the nanoparticles loaded into cells, i.e. could be useful in drug delivery of chemotherapy or hyperthermia agents.

Figure 3 shows the MI curves for different amplitudes of the driving currents for as-quenched Fig 3. (a) and gold covered CoFeCrSiB amorphous ribbons Fig 3. (b). The dependence of the MI ratio maximum on the amplitude of the driving current is rather weak for the amplitudes under consideration. Therefore an intermediate current of 30 mA was selected for testing with biological samples. The MI effect is noticeably decreased for the gold covered ribbon. It could be attributed to the non-uniform current distribution The MI of a uniform ferromagnetic conductor is based on the dependence of skin penetration depth (δ) on transverse magnetic permeability [12]. In the case of gold plate the skin penetration depth at 5 MHz can be estimated as 35 μm (comparing with 12 μm for amorphous ribbon of selected composition). In the other words, in the case of the covered ribbons the current density on the amorphous ribbons (which contributes to MI effect) could be less than that of the uncovered ribbons because the increased fraction of the current flows through the gold layer. As a consequence the MI effect decreases for the covered ribbons. Although both the total value of the effect and MI sensitivity with respect to a magnetic field (about 43%/Oe for uncovered and 35%/Oe for gold covered cases) are reduced for the gold covered ribbons in spite of this the MI effect sensitivity is still high. The gold covering can be

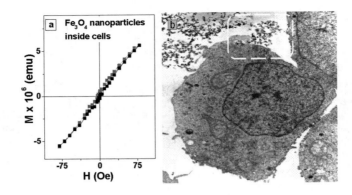

FIGURE 4. a) Longitudinal hysteresis loop of Fe_3O_4 introduced into HEK 293 cells by uptake events, measured by SQUID; b) TEM microscopy: the progress of Fe_3O_4 magnetic nanoparticles uptake inside HEK 293 cell where the dashed line indicates the particles introducing it on the cells.

considered as an improvement for certain applications even at the price of a decrease in sensitivity because the MI response of the protected sensitive element does not depend on the biocorrosion processes. Another advantage - gold is biocompatible and can be used to help the formation monolayers of probe molecules for binding assays in certain regimes of biosensing [13]. The comparison of the shape of the hysteresis loops and MI curves (Fig 3) reveals that the magnetic properties of uncovered and gold covered ribbons are similar. The effective magnetic permeability is longitudinal in both cases and the MI effect is close to one-peak response with specific split into two-peak shape in small fields, corresponding to surface anisotropy contribution of as-quenched ribbons [5].

Magnetic nanoparticles are widely employed in biomedical applications and biosensing [14]. In particular, iron oxide particles are used in magnetically targeted drug delivery because they have the advantage of injectability and high level accumulation in the target tissue [15]. In drug delivery of cancer combating agents one can utilize the ability of Fe_3O_4 surface modified superparamagnetic nanoparticles to turn cancer cells into mini magnets by accumulation of the nanoparticles inside them [15]. The hysteresis loops of the Fe_3O_4 nanoparticles loaded into cells can be seen in Fig. 4(a). Close to zero coercivity and remanence as well as S-like shape of the *M(H)* curve indicated a superparamagnetic dominating contribution.

For a biosensor prototype testing, the applied magnetic field dependence of MI ratios of non-covered and gold covered amorphous ribbons was measured in a decreasing magnetic field in two regimes: for a cell sample containing nanoparticles and without them. The first test was provided with sensitive element immersed in solutions with cells and without them. These two responses were identical with accuracy better than 0.5%, confirming the stability to the chemically aggressive medium of the bath. Figure 5 shows the results of the difference of MI responses corresponding to cells without particles and with nanoparticles introduced into cells by up-take events. A clear difference between MI responses was observed for cells without particles and with Fe_3O_4 nanoparticles introduced into cells. This difference

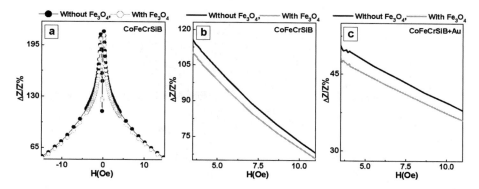

FIGURE 5. MI of CoFeCrSiB rapidly amorphous ribbons in a decreasing field for alternating current intensity of 30 mA in presence of nanoparticles of Fe_3O_4 and without it (a and b). c) MI of composite of CoFeCrSiB and gold in presence of nanoparticles of Fe_3O_4 and without it.

between two responses could be interpreted as a result of disturbances of the applied magnetic field by fringe fields of magnetic particles introduced into cells.

In summary, MI biosensors for the detection of the presence of Fe_3O_4 magnetic nanoparticles after cellular uptake has been presented and tested for a model HEK-293 cells system. The MI response of the sensitive element is clearly dependent on the presence of magnetic nanoparticles both for non-covered and gold covered ribbons. Although the sensitivity of the prototype with gold layer is reduced, this configuration has the advantage of the biocompatibility and the stability of the response as well as gives one the chance to use the protecting layer for self assembling process. The results confirm the possibility of designing MI biosensor for biomedical applications to evaluate the level of the intracellular uptake of magnetic nanoparticles which can be used as biomolecular labels.

ACKNOWLEDGMENTS

This work has been supported in Spanish by a "Ramon y Cajal" grant of Spanish MEC and UPV-EHU, by the grants MAT 2005-06806-C04-03 and MAT-2003-06407 of Spanish MEC. The authors are deeply grateful to M. A. Cerdeira, J. A. Garcia, H. Srikanth and J. Gass for special support.

REFERENCES

1. D.R. Baselt, G.U. Lee, M. Natesan, S.W. Metzger, P.E. Sheehan and R.J. Colton, *Biosens. Bioelectron.* **13**, 731-739 (1998).
2. M.M. Miller, G.A. Prinz, S.F. Cheng and S. Bounnak, *Appl. Phys. Lett.* **81**, 2211-2213 (2002).
3. G.V. Kurlyandskaya, M.L. Sánchez, B. Hernando, V.M. Prida, P. Gorria and M. Tejedor, *Appl. Phys. Lett.* **82**, 3053-3055 (2003).
4. N. Bouropoulos, D. Kouzoudis and C. Grimes, *Sens. Actuators B* **109**, 227-232 (2005).
5. G.V. Kurlyandskaya, V. Fal Miyar, A. Saad, E. Asua and J. Rodríguez, *J. Appl. Phys.* **101**, 054505 1-9 (2007).
6. G.V. Kurlyandskaya and V. Levit, Biosens. Biolelectr. **20**, 1611 (2005).

7. R.S. Beach and A.E. Berkowitz, *Appl. Phys. Lett.* **64**, 3652-3654 (1994).
8. M. Megens, F. Theije, B. Boer and F. Gaal, *J. Appl. Phys.* **102**, 014507 1-5 (2007).
9. F.L. Graham, J. Smiley, W.C. Russell and R. Nairn, *J. Gen. Virol.* **36**, 59-74 (1977).
10. Clontech Laboratories, California U.S.A. http://www.clontech.com. (2001)
11. M. Mandal, S. Kundu, S.K. Ghosh, S. Panigrahi, T.K. Sau, S.M. Yusuf and T. Pal, *J. Colloid Interface Sci.* **286**, 187-194 (2005).
12. L.D. Landau and E.M. Lifshitz, "Electrodynamics of Continuous Media", Oxford: Pergamon, 1975, pp. 195.
13. I.O. K'Owino, S.K. Mwilu and O.A. Sadik, *Anal. Biochem.* **369**, 8–17 (2007).
14. Q.A. Pankhurst, J. Connolly, S.K. Jones and J. Dobson, *J. Phys. D: Appl. Phys.* **36**, R167-R181 (2003).
15. A.S. Lübbe, C. Alexiou and C. Bergeman, *J. Surgical Research* **95**, 200-206 (2001).

Residence Times Difference Fluxgate Magnetometer for Magnetic Biosensing

B. Andò*, A. Ascia*, S. Baglio*, A.R. Bulsara+, V. In+, N. Pitrone*, and C. Trigona*

*Dipartimento di Ingegneria Elettrica, Elettronica e dei Sistemi, University of Catania, Via Andrea Doria 6, 95125, Catania, Italy
+Space and Naval Warfare Systems Center, Code 2363, 49590 Lassing Road, San Diego, CA 92152-6147, USA

Abstract. In this paper we discuss the performance of Residence Times Difference fluxgate magnetometers when used in magnetic immuno-assay applications. Two approaches for magnetic immuno-assay techniques are discussed here, both devoted to magnetic bead detection: the first is oriented to detect beads in liquid solutions while the second focuses on the detection of small spots (diameter on the order of 100 μm) of magnetic beads. Two different fluxgate magnetometers are used here: a "FR4 Fluxgate", using a Metglas Magnetic Alloy 2714 As Cast (Cobalt based) core, and a "FeSiB Amorphous magnetic microwire fluxgate", having a magnetic core with a diameter of 100 μm. Both sensors feature high sensitivity to weak static magnetic fields and are seen to perform well in detecting magnetic beads in liquid solutions; the microwire based device also allows for measurements with a high spatial resolution.

Keywords: RTD-fluxgate, FR4-fluxgate magnetometer, Microwire-fluxgate magnetometer, FeSiB, Magnetic beads.

PACS: 07.55.Ge

INTRODUCTION

Magnetic bioassaying is gaining great interest due to recent advances in magnetic materials and in highly sensitive detection techniques [1]. The main idea behind these techniques is the possibility of labelling target bio-entities with magnetic particles that are used to perform direct or indirect tasks such as targeted delivery or labelling.

A large ensemble of magnetic particles is available for these applications; the particles differ in magnetic properties, sizes and biocompatibility. Labelling is made possible, usually, through coating of the magnetic particle surfaces with biocompatible molecules providing a link between the particle and the target. A large range of coatings is commercially available, in order to suit targets including sites on biological macromolecules such as DNA and antigens on the surfaces of cells. For example, magnetic particles coated with immunospecific agents have been successfully bound to red blood cells [2, 3], lung cancer cells [4], bacteria [5] and other target entities [6, 7]. Several applications in the field of magnetic bioassaying have been, recently, identified; they include magnetic separation, drug delivery, hyperthermia treatments, magnetic resonance imaging (MRI), and magnetic labelling.

CP1025, *Biomagnetism and Magnetic Biosystems Based on Molecular Recognition Processes*
edited by J. A. C. Bland and A. Ionescu
© 2008 American Institute of Physics 978-0-7354-0547-9/08/$23.00

Magnetic labelling techniques can be used to identify target entities adopting a remote sensing approach. The most common techniques are AC susceptometry, SQUID magnetometry, or giant magnetoresistance (GMR) sensors [8, 9, 10]; although the latter devices are less sensitive than SQUIDs, the extreme proximity of the sensor to the inspected entities tagged by magnetic beads can dramatically boost its performance and offset the lack of sensitivity.

Fluxgate magnetometers represent an alternative solution to sense weak magnetic fields, or perturbations thereof, at room temperature. Recently, the authors have proposed Residence Times Difference (RTD) fluxgate as competitive devices to the traditional second harmonic architectures [11, 12]. Low cost, small dimension, high sensitivity, low noise floor, low power consumption and an intrinsic digital form of the output signal are the main advantages afforded by this innovative readout strategy.

In this paper, the use of RTD-fluxgates as candidates for magnetic bead detection in immuno-bioassay applications is discussed. First, the RTD-fluxgate working principle is discussed and the analytical model presented. In the context of innovative architecture of the RTD-Fluxgate magnetometer to detect static magnetic field, two technological prototypes are here presented: 1) FR4-fluxgate magnetometer (multilayer PCB fluxgate), and 2) Microwire-fluxgate magnetometer. The two sensors have active areas of different sizes (a few millimetres for the FR4 device and 100 μm for the microwire sensor) so that they address different applications.

Background: The RTD-Fluxgate Magnetometer

In this section a brief overview of the RTD-fluxgate magnetometer is presented; more details can be found in [11, 12].

An RTD Fluxgate is based on a two-coils structure (a primary coil and a secondary coil) wound around a single ferromagnetic core having a hysteretic input-output characteristic as shown in Figure 1. A periodic driving current, I_e, is forced in the primary coil and generates a periodic magnetic field, H_e. A target field H_x is applied in the same direction of H_e.

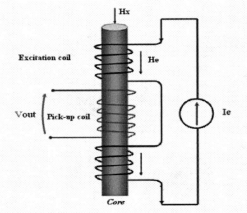

FIGURE 1. The RTD-fluxgate sensor structure.

We assume the ferromagnetic core has two commutation thresholds and a two state output, whose behaviour is predicated by micromagnetic phenomena and is usually obtained via mean-field approximations to the collective motion of the core domain walls. The potential function $U(x)$ is given by:

$$U(x) = \frac{x^2}{2} - \frac{1}{c}\ln\cosh[c(x + H_e(t) + H_x)]$$

(1)

The magnetization x of the core is governed by the excitation field, H_e, produced in the primary coil, and by the bistable potential energy function $U(x)$, which underpins the crossing mechanism between the two steady magnetization states of the magnetic core. In order to reverse the core magnetization (from one steady state to the other one), the driving field (H_e) must cross the switching thresholds of the magnetic core. The model that describes the dynamical behaviour of the magnetization $x(t)$ in the ferromagnetic core via the dynamical system is, then, expressed via the equation [13]:

$$\tau\frac{dx}{dt} = -x + \tanh\left[\frac{x + H_e(t) + H_x}{K}\right] \equiv -\frac{\partial U(x,t)}{\partial x},$$

(2)

where H_e and H_x represent the excitation and external magnetic fields, respectively, τ is the system time constant and K a dimensionless, temperature dependent, control parameter. In Figure 2 the qualitative relation between the potential function and the hysteresis loop is shown. In presence of a ferromagnetic core, when the excitation field exceeds the positive and negative coercive fields H_c and $-H_c$, the magnetization x evolves alternatively from the positive saturation state to the negative saturation state and vice versa.

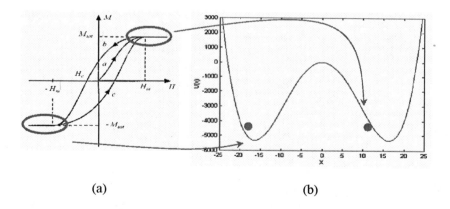

(a)   (b)

FIGURE 2. Relation between the positive and negative stable states of the hysteresis cycle (a) and the potential energy function U(x) (b). In (b) normalized units are used.

141

Assuming a magnetic core having a sharp hysteretic characteristic, the switch between the two stable states of the magnetization occurs nearly instantaneously when the external magnetic field exceeds the coercive field level H_c. Under this hypothesis the device operates as a static hysteretic nonlinearity (e.g a Schmitt Trigger).

Under these assumptions it is possible to define the residence time T^+ as the time interval between the crossing of the upper H_c level (at time t_1, Figure 3) and the successive crossing of the lower H_c level (at time t_2); defining the residence time T^- as the time interval between the crossing of the lower H_c level (at time t_2) and the upper level in the next period of the bias signal (at time t_3). The Residence Times Difference is defined as $RTD = T^+ - T^-$, it is null in the absence of external target magnetic field.

In the case of a time-periodic excitation having amplitude large enough to cause switching between the steady states and in the absence of any target field, the hysteresis loop (or the underlying potential energy function $U(x)$) is symmetric and two identical Residence Times are obtained.

The presence of a target DC field, H_x, leads to a skewing of the potential function with a direct effect on the Residence Times: the RTD: for a nonzero target signal, the RTD is non-zero.

The RTD is evaluated by using the voltage signal v_{out} picked-up by the secondary coil, this voltage is proportional to the first derivative of the magnetization and, dealing with a soft magnetic material which shows a sharp hysteresis loop, it appears

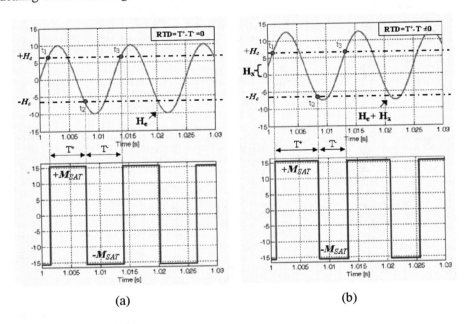

(a)   (b)

FIGURE 3. Temporal evolution of the bias field and of the magnetization in absence (a) and in presence of an external magnetic field (H_x).

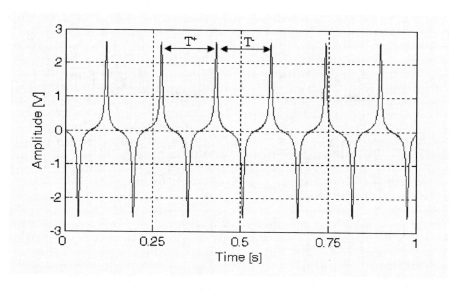

FIGURE 4. A typical pick-up coil output voltage signal.

as a sequence of sharp pulses that arise at the instants of commutation between one saturation state and the other. The estimation of the relative peak position leads to the evaluation of the RTD.

In Figure 4 a typical output voltage time evolution is shown just to better explain the above stated readout strategy.

Two RTD-Fluxgate Prototypes

In this section two different realizations of RTD-fluxgate magnetometers are presented. The former technological approach is based on the use of a 100 μm thick ferromagnetic ribbon embedded between two FR4 layers for a total thickness of 1.6 mm, while the latter system is based on the use of 100 μm ferromagnetic amorphous microwires.

1. The RTD-fluxgate adopted in the experimental set-up for the detection of magnetic beads in solution is realized in Printed Circuit Board (PCB) technology. The Magnetic Alloy 2714 As Cast (Cobalt based), by Metglas® [14], is chosen as magnetic core layer due to its suitable hysteretic characteristic which affords the use of a readout strategy based on the estimation of the RTD. In the process, a patterned Metglas foil (wet etching is adopted for the Metglas patterning) is embedded between two FR4 PCB layers. A simplified process description can be summarized by the following steps: the patterned Metglas is aligned with respect to the two FR4 layers, the patterned metal layers are aligned to the sensor structure according to the layout design, and the whole assembly is pressed, while heating the whole system up to 200°C. Finally the vias are formed between the lower and upper layers to permit completion of the coil windings.

143

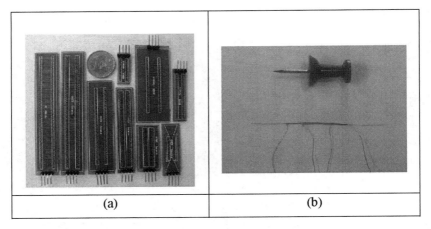

| (a) | (b) |

FIGURE 5. a) Image of a set of PCB integrated fluxgate sensors. b) RTD-fluxgate magnetometer in "wire core" technology (microwire-fluxgate magnetometer).

Figure 5(a) shows a set of RTD-fluxgates developed in PCB technology with an embedded foil of Metglas [15, 16]. The device characterization has been exhaustively addressed in [17-19].

2. The "wire core" materials have a 100 μm magnetic structure diameter [20] (FeSiB Amorphous magnetic microwire). The RTD fluxgate exploiting the "wire core" can be used to measure spot magnetic fields in a particular direction, in fact it allows for pointing at very small targets. A simplified process description can be summarized by the following steps: a two coil structure (primary coils and secondary coils) is wound around a (glass) support, and 100 μm ferromagnetic amorphous wire is inserted in the (hollow) glass support. Figure 5(b) shows RTD-fluxgate magnetometer in "wire core" technology.

Magnetic Beads Detection with the RTD-Fluxgate

Bio-applications of magnetic beads require an optimized sensing architecture to detect small concentration of magnetic particles. Actually, both high sensitivity and high spatial-resolution are required depending on the specific application.

Bio-assay and off-line measurements of magnetically labelled species contained in spotted samples are two examples where high performing magnetic detectors are mandatory.

In the next sections two examples of magnetic beads detection by RTD-fluxgate are briefly introduced to highlight performances of both the sensing architecture developed and the detection strategy adopted.

Detection of Magnetic Beads in Solution

In this section the experimental characterization of a PCB fluxgate prototype for estimation of magnetic beads in aqueous solution is discussed. In particular, magnetic beads CM-10-10 from Spherotech [21] with paramagnetic properties and spherical structure are used.

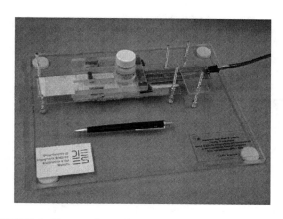

FIGURE 6. The experimental set-up for magnetic beads detection.

The task to be accomplished within this experiment is the detection of micrometer sized magnetic particles. Particles (spherical in shape, uniform in size and paramagnetic) are prepared by coating a layer of magnetite and polystyrene onto a monodispersed polystyrene core. They become non-magnetic when removed from a magnetic field, and do not retain any detectable magnetism even after repeated exposure to a strong magnetic field. The experimental set-up is shown in Figure 6.

It consists of a RTD-fluxgate magnetometer, operated with a sinusoidal bias current at 80 Hz, a permanent magnet to polarize the magnetic particles and a glass pipe containing the suspension provided by Spherotech.

The experiment has been conduced by observing the fluxgate output for varying values of two parameters: the distance between the beads and the magnetometer, and the number of beads. RTDs estimated through the experimental set-up are given in Figures 7(a) and (b), where the evolution of the RTD for six different values of the target distances and two different quantities of beads is reported.

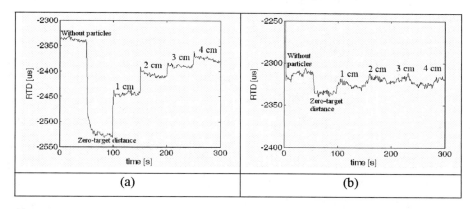

FIGURE 7. a) RTD evolution for six different values of target distance and 10^{10} particles. b) RTD evolution for six different values of target distance and 10^8 particles.

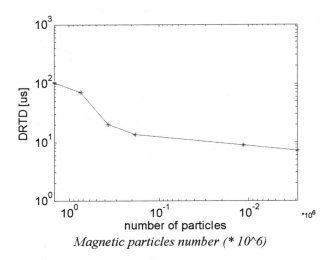

Magnetic particles number (* 10^6)

FIGURE 8. RTD variation in the case of zero target distance as function of the beads number (both axes are in logarithmic scales).

Each plot reproduces the change in the RTD when the magnetic sample is positioned at increasing distance from the measuring device, therefore simulating the reduction of the target signal amplitude.

The RTD variation in the case of zero target distance is given in Figure 8, where we plot the DRTD (Variation of Residence Times Difference) evolution as a function of the number of magnetic beads in the solution.

The smallest detectable bead number has been estimated by taking into account the limitations imposed by the intrinsic noise floor on the readout signal. During the experiments a noise level of 5 μs (in the RTD) has been evaluated by measuring the fluctuations of the output signal in the absence of any target. The Figure 8 shows a minimum detectable number of beads approximately equal to $3*10^3$.

Detection of Spotted Magnetic Beads

In this section the experimental characterization of a "wire core" fluxgate prototype for estimation/characterization of spotted magnetic beads is discussed. "Spots" are obtained by the deposition of controlled quantities of fluid, containing beads with a known concentration, over a given surface. The microwire fluxgate magnetometer can be used to detect magnetic beads (CM-10-10) immobilized onto 100 μm-thick glass surface.

The experimental set-up consists of a RTD-fluxgate magnetometer with ferromagnetic amorphous wire, excited with the same sinusoidal bias frequency of the PCB prototype. To polarize the magnetic particles in the deposited spot, two permanent magnets are used. The magnetic wire is then bent to form a magnetic circuit that closes through the magnetized beads deposited on the spot; the primary and secondary coils are then wound over the magnetic wire to form the fluxgate.

The experimental set-up is shown in Figure 9(b). The experiment has been conduced by taking several observations of the fluxgate output for various spot sizes (obtained via the deposition of different known fluid volumes ranging from 0.5 µl to 4 µl, Figure 9(a)). Each spot is maintained in the measuring position for 300 seconds. Figure 9(c) shows the averaged RTD, after removing the transient response, for five magnetic spots of different size (corresponding to 0.5 µl, 1 µl, 2 µl, 3 µl and 4 µl respectively). The circled point shown in Figure 9(c) indicates the device response in the absence of any target spot.

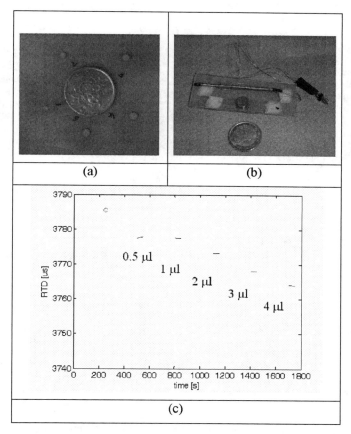

FIGURE 9. a) Magnetic beads (CM-10-10) immobilized onto 100 µm-thick glass surface. b) The experimental set-up for spotted magnetic beads detection. c) RTD evolution for the microwire fluxgate magnetometer; the graphs reports the experimental results obtained by applying the sensor to target with an increasing number of magnetic particles deposited.

CONCLUSIONS

The results reported in this paper demonstrate the feasibility of using RTD-fluxgate magnetometers in magnetic immuno-assay applications. Two sensor structures with active areas of different sizes (some millimetres for the FR4 device and 100 μm for the wire core sensor) have been proposed for applications where the target size changes from the case of particles dispersed in liquids contained into a pipe to particles spotted over a micrometric surface.

REFERENCES

1. Q. A. Pankhurst, J. Connolly, S. K. Jones and J. Dobson, Applications of magnetic nanoparticles in biomedicine, Journal of Physics D: Applied Physics **36**, issue 13, R167–R181(1) (2003).
2. R. S. Molday and D. MacKenzie, Immunospecific ferromagnetic iron–dextran reagents for the labeling and magnetic separation of cells, *Journal of Immunological Methods* **52**, 353–367 (1982).
3. A. Tibbe, B. De Grooth, J. Greve, P. Liberti, G. Dolan and L. Terstappen, Optical tracking and detection of immunomagnetically selected and aligned cells, *Nature Biotechnology* **17**, 1210–1213 (1999).
4. B. Y. Kularatne, P. Lorigan, S. Browne, S. K. Suvarna, M. O. Smith and J. Lawry, Monitoring tumour cells in the peripheral blood of small cell lung cancer patients, *Cytometry* .**50**, issue 3, 160–167 (2002).
5. S. Morisada, N. Miyata and K. Iwahori, Immunomagnetic separation of scum-forming bacteria using polyclonal antibody that recognizes mycolic acids, *Journal of Microbiological Methods* **51**, issue 3, 141–148 (2002).
6. R. E. Zigeuner, R. Riesenberg, H. Pohla, A. Hofstetter and R. Oberneder, Isolation of circulating cancer cells from whole blood by immunomagnetic cell enrichment and unenriched immunocytochemistry *in vitro*, *Journal of Urology* **169**, issue 2, 701–705 (2003).
7. C. V. Mura, M. I. Becker, A. Orellana and D. Wolff, Immunopurifcation of Golgi vesicles by magnetic sorting, *Journal of Immunological Methods* **260**, issue 9, 263–271(2002).
8. C. B. Kriz, K. Radevik and D. Kriz, Magnetic permeability measurements in bioanalysis and biosensors, Analytical Chemistry **68**, issue 11, 1966-1970 (1996).
9. R. Kotiz, H. Matz, L. Trahms et al., SQUID based remanence measurements for immunoassays, IEEE Transactions on Applied Superconductivity **7** (2), 3678-3681 (1997).
10. M. M. Miller, P. E. Sheehan, R. L. Edelstein, C. R. Tamanaha, L. Zhong, S. Bounnak, L. J. Whitman and R. J. Colton, A DNA array sensor utilizing magnetic microbeads and magnetoelectronic detection, Journal of Magnetism and Magnetic Materials **225**, issue 1-2, 138-144 (2001).
11. B. Andò, S. Baglio, A. Bulsara and V. Sacco, Theoretical and experimental investigations on residence times difference fluxgate magnetometers, *Measurement (Elsevier)* **38** (2), 89-112 (2005).
12. B. Andò, S. Baglio, A. Bulsara and V. Sacco, "Residence Times Difference" Fluxgate Magnetometers, *Sensors Journal, IEEE* **5** (5), 895-904 (2005).
13. H. E. Stanley, *Introduction to Phase Transitions and Critical Phenomena*, Oxford University Press, Oxford and New York, 1971.
14. www.metglas.com.
15. B. Andò, S. Baglio, V. Caruso, V. Sacco and A. Bulsara, Multilayer based technology to build rtd fluxgate magnetometer, IFSA, *Sensors and Transducers Magazine* **65**, issue 3, 509-514 (2006).
16. B. Andò, S. Baglio, V. Caruso and V. Sacco, "Investigate the optimal geometry to minimize the demagnetizing effect in RTD-Fluxgate" in *IEEE IMTC 2006*, Proceedings of the IEEE-Instrumentation and Measurement Technology Conference, 2006, 2175-2178.
17. B. Andò, S. Baglio, V. Sacco, A.R. Bulsara and V. In, "PCB Fluxgate Magnetometers with a Residence Times Difference (RTD) Readout Strategy: The Effects of Noise" in *IEEE Transactions on Instrumentation and Measurement* **57**, 19-24 (2008).

18. B. Andò, A. Ascia, S. Baglio, A. R. Bulsara, J. D. Neff and V. In, Towards the Optimal Reading of RTD Fluxgate, *Sensors and Actuators A: Physical* **142**, 73-79 (2008).
19. B. Andò, A. Ascia, S. Baglio, A.R. Bulsara, C. Trigona and V. In., "RTD Fluxgate performance for application in magnetic label-based bioassay: preliminary results" in Proceedings of the 28th Conference of the IEEE – EMBS, 2006, pp. 2058-2061.
20. H. Chiriac, J. Yamasaki, T. A. Ovari and M. Takajo, Magnetic domain structure in amorphous glass-covered wires with positive magnetostriction, *IEEE Transactions on Magnetism* **3**, issue 5, 3901-3903 (1999).
21. www.spherotech.com

Integrated Spintronic Platforms for Biomolecular Recognition Detection

V.C. Martins[a, b], F.A. Cardoso[a, c], J. Loureiro[a, c], M. Mercier[a, c], J. Germano[c, d], S. Cardoso[a, c], R. Ferreira[a, c], L.P. Fonseca[b], L. Sousa[c, d], M.S. Piedade[c, d] and P.P. Freitas[a, c]

[a]INESC-Microsystems and Nanotechnologies, Rua Alves Redol 9, 1000-029 Lisbon, Portugal.
[b]Center for Biological and Chemical Engineering, Instituto Superior Técnico, Av. Rovisco Pais 1049-001 Lisboa, Portugal.
[c]Instituto Superior Técnico, Avenida Rovisco Pais, 1000-029 Lisbon, Portugal.
[d]INESC-Investigação e Desenvolvimento, Rua Alves Redol 9, 1000-029 Lisbon, Portugal.

Abstract. This paper covers recent developments in magnetoresistive based biochip platforms fabricated at INESC-MN, and their application to the detection and quantification of pathogenic waterborn microorganisms in water samples for human consumption. Such platforms are intended to give response to the increasing concern related to microbial contaminated water sources. The presented results concern the development of biological active DNA chips and protein chips and the demonstration of the detection capability of the present platforms. Two platforms are described, one including spintronic sensors only (spin-valve based or magnetic tunnel junction based), and the other, a fully scalable platform where each probe site consists of a MTJ in series with a thin film diode (TFD). Two microfluidic systems are described, for cell separation and concentration, and finally, the read out and control integrated electronics are described, allowing the realization of bioassays with a portable point of care unit. The present platforms already allow the detection of complementary biomolecular target recognition with 1 pM concentration.

Keywords: Magnetoresistive sensors, Biochips, Biosensors, Surface biochemistry, Biomolecular recognition, Magnetic labels.
PACS: 85.90.+h, 87.80.Fe, 87.85.fk

1. INTRODUCTION

Magnetoresistive sensors are widely used in the data storage industry as read heads in hard disk drives [1]. Recently their application as magnetoresistive biosensors or biochips for the detection of biomolecular recognition events has also emerged as a promising technology [2]. MR biosensors take advantage of the high magnetic field sensitivity of the magnetoresistive transducers to detect the fringe field of magnetic microspheres or nanoparticles used as biomolecular target markers. An increasing number of scientific groups are dedicated to investigate MR biosensors [3-10] as its promised potential is being unveiled in terms of sensitivity [11], reliability as well as integrability [12], versatility and more recently scalability [13].

CP1025, *Biomagnetism and Magnetic Biosystems Based on Molecular Recognition Processes*
edited by J. A. C. Bland and A. Ionescu
© 2008 American Institute of Physics 978-0-7354-0547-9/08/$23.00

INESC-MN and its collaborators are pursuing the development of an integrated platform where the magnetic sensing unit, the biochemical interface (functionalized organic/inorganic surface), the macro/micro-fluidic system (sample preparation, cell concentrator, fluid transport), and the electronic data read-out and automation are combined in an easy-to-use and portable platform. This paper presents an overview of the latest achievements on each of these different issues. Two types of sensing units (spin valves (SV) and magnetic tunnel junctions (MTJ)) and chip architectures will be presented and compared in terms of sensitivity and signal-to-noise ratio [14]. The surface biochemistry in use will be described in detail and the biological detection limit of the system described, for detection of the hybridization of complementary 20-mer single stranded DNA sequences. A particular application is being targeted on the evaluation of microbiological quality of water for human consumption through the detection of waterborne pathogen microorganisms. Two different strategies of biomolecular recognition events, involving DNA oligonucleotide sequences and antibodies as recognition agents, are under development for the quantitative determination of *Escherichia coli* and *Salmonella sp.* as model microorganisms in water samples. Our goal is to simultaneously develop DNA and protein chips seeking the complementarities and validation of both strategies. Biochemical modification of this type of chip surfaces is not straightforward and many challenges were faced in order to optimize chemical procedures and achieve a stable, reproducible and bioactive array of probes (DNA single strands and antibodies). With the presently optimized surface derivatization, the DNA chip is able to detect as low as 1 pM of target DNA strands. Protein chips still face some degree of non-specific binding events. Further improvements in the detection limit have already been demonstrated by the use of magnetic attraction and focusing on sensor site of magnetically labelled targets [15-17].

Regarding the protein chips, simulated *Salmonella* suspensions were used to perform a "sandwich"-type immunoassay, where general difficulties were identified and are being addressed, *e.g.* procedure optimization, non-specific adsorption of microbial cells and washing reproducibility issues. In case of real water samples and whole microbial cells detection, an initial volume of not less than 100 mL is regulated to be analyzed [18]. In that case a macro/micro-fluidic interface is required in order to allow real time analysis in such a compact device. Addressing this issue, preliminary results on the design and fabrication of a microbial cell micro-concentrator based on magnetic field for integration on the biochip microfluidic platform are shown.

Additionally, latest achievements on signal acquisition and processing with relevant electronic noise reduction will be discussed.

In sum, the magnetoresistive biochip arrays, being fabricated at INESC-MN, are successfully being integrated into a fully portable, miniaturized platform, including all control and read out electronics, as well as the microfluidic platform for sample handling, including concentration and separation.

2. SPIN-VALVE-BASED BIOCHIP

2.1. Sensor and Chip Microfabrication

The biochip comprises 24 U-shaped spin valve sensors arranged in 4 parallel rows, each sensor surrounded by a U-shaped current line. The clean room microfabrication process starts with a 3 inch silicon wafer as a substrate. After an initial passivation layer of 500 Å of aluminum oxide (Al_2O_3), the spin valve metallic multi-layer structure was deposited using an ion beam deposition system. The multi-layer structure consisted of Ta 15 Å/ NiFe 30 Å/ CoFe 25 Å/ Cu 21 Å/ CoFe 25 Å/ MnIr 80 Å/ Ta 20 Å/ TiW(N) 150 Å. The pinned and free layer easy axis are at 90 degrees from each other and the pinned layer easy axis is transverse to the sample height (easy axis defined by the applied in-plane field [-40 Oe] during deposition). The sensors were defined by direct write laser photolithography and ion milling, resulting in U-shaped sensors of dimensions 2.5 (height) x 130 (length) μm^2 with an average magnetoresistance of 7%. Sensor leads and U-shaped current lines surrounding each sensor were made of 3000 Å thick aluminum deposited by magnetron sputtering and completed by lift-off. In particular, metallic current lines are tailored such that magnetic labels and magnetically labeled targets are focused over the sensing area (Fig. 1). The whole structure was covered with a double sputtered passivation layer, respectively, 1000 Å of Al_2O_3 and 2000 Å of silicon dioxide (SiO_2). Contact pads around the edges of the chip were opened by lift-off to allow the wire-bonding to a 40-pin chip-carrier. Additionally, small pads of dimension 40 x 13 μm^2 consisting in Ti 50Å/ Au 200Å were RF sputtered and patterned by lift-off precisely on top of sensor sites (Fig. 1).

After wire-bonding, the wires were protected from the external environment by a layer of silicon gel (Elastosil E41) deposited directly over the wires all around the chip, simultaneously defining the limits of an open chamber for further confinement of biochemical reactional solutions.

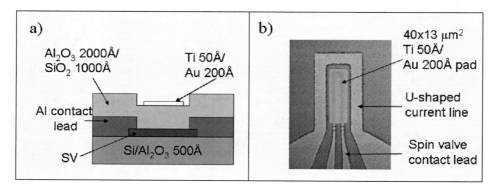

FIGURE 1. Crossectional (a) and top (b) view of a spin-valve chip sensing site showing the sensor, focusing current line, passivation layers and gold pad.

2.2. Biochemical Reagents

All reagents used in the biochemical tests were of analytical grade. Sodium cholate hydrate, 3-(2-pyridyldithio) propionic acid N-hydroxysuccinimide ester (SPDP) and DL-dithiothreitol (DTT) were obtained from Sigma. Bovine Serum Albumin (BSA) fraction V was purchased from Merck, Tween20® was from Promega and polyethyleneglycol, HO-PEG-SH, of 3kDa ordered from RAPP Polymere (Tübingen, Germany). Single stranded customized 20 and 35-mer oligonucleotides (ssDNA) were synthesized by MWG-Biotech (Ebersberg, Germany), encoding for a sequence of the conservative region of the 16S rDNA from *E. coli*. Their designation, base sequences and modifications are as follows: probe, Agk11A2, 5' SH – TTT TTT TTT TTT TTT ACA CGG TCC AGA CTC CTA CG- 3'; complementary target, Agk11F, 5' biotin - CGT AGG AGT CTG GAC CGT GT- 3'; and non-complementary target Agk11NC30, 5' biotin - TCA ATG GAG CTA CTC ATA GC - 3'.

Polyclonal antibodies, unconjugated and biotinylated, from rabbit anti *Salmonella sp.* were obtained from Biodesign International (Ohio, USA) and *Salmonella thiphimurium* cells were a kind gift from the Water Analysis Laboratory from Instituto Superior Técnico, Lisbon, Portugal.

Nanomag®-D, 250 nm, and Micromer®-D, 3 μm, magnetic particles, modified with streptavidin, were obtained from Micromod, Germany.

TRIS-EDTA buffer (TE-HCl) was prepared from 10 mM TRIS with addition of 1 mM EDTA, 0.1 M K_2HPO_4 (unless otherwise stated), and 1 M HCl to adjust the pH to 7.4.

Phosphate buffer (PB), 100 mM, consisted of a combination of monobasic sodium phosphate (NaH_2PO_4) and dibasic sodium phosphate (Na_2HPO_4), while the pH was adjusted to 7.4. Phosphate saline buffer (PBS) was prepared from a stock solution 10x concentrated which consisted of a combination of Na_2HPO_4, KH_2PO_4, NaCl and KCl; pH 7.4 adjusted with HCl.

2.3. Surface Biochemistry

Despite being known as physically robust, complementary metal-oxide-semiconductor (CMOS) devices are in great extent perishable when in contact to most of the well-established solid surface chemistries for biosensing applications. Before we succeeded in the realization of biomolecular recognition detection assays, a great effort has been made in order to achieve an efficient, stable and reproducible surface chemistry.

An important parameter to evaluate, when implementing a surface chemistry for the development of a biochip, is the harshness of its solutions, namely in terms of pH, salt concentration, ionic strength, organic moieties and temperature. Simultaneously, a compromise between the thickness of the passivation layer used and the efficiency and the harshness of the chosen protocol should be taken into account. The thiol-gold biochemistry is in particular a very demanding protocol. It is characterized by the usage of extremely harsh cleaning solutions, *e.g.* Piranha (70% (v/v) sulfuric acid/ 30% (v/v) hydrogen peroxide) [19], strong acids (HCl) or alkaline solutions (NH_4OH, NaOH) [20], in order to keep the gold surface free from organic contaminants.

Another critical factor is the high salt concentration, up to 1 M, often reported for thiol-modified molecules immobilization in order to achieve a more stable sulphur-gold linkage [19, 21]. During the optimization of a non-damaging thiol-gold chemistry protocol two main parameters were evaluated: I.) the cleaning procedure before probe immobilization and II.) the salt concentration effect during immobilization, blocking, hybridization and magnetic labeling steps. The optimization of surface chemistry protocols was performed on dummy gold substrates to avoid wasting processed chips and chip-carriers.

In preparation for thiolated probes immobilization, the chips or gold substrates went through an optimized and meticulous mild cleaning procedure. The substrates were dipped into hot (80°C) Microstrip 2001 for 15 min, copiously rinsed with acetone, isopropyl alcohol (IPA), ultra-pure milli-Q grade water, and blown dried with a nitrogen stream. To completely remove the traces of the photoresist polymer and other organic contaminants the gold surfaces were exposed to oxygen plasma for 60 sec at 100 W, 100 sccm O_2 and 300 mT in the load-lock of a Plasmatherm machine. Immediately afterwards the chips were submerged in a cholic acid solution (2% w/v) overnight and finally rinsed with water and blown dried under a nitrogen stream. Then, in the case of gold substrates the thiolated probes were manually spotted through a procedure depicted in Figure 2. For processed and encapsulated chips they were simply covered with a 50 μL droplet of DNA solution. The discrete multi-spotting procedure was adopted as a strategy to easily distinguish between specifically bound species (*in-spot*) and non-specifically adsorbed species or background (*out-spot*). Control experiments for both DNA and antibody arrays were performed where a non-complimentary target or a non-specific biorecognition element was respectively used to assess for non-specific recognition background signals.

The whole process from immobilization through blocking, biorecognition and magnetic labeling steps was performed inside a Petri dish in a humid atmosphere to prevent evaporation.

At the end, after magnetic labeling and washing steps, optical microscopic images were taken using a CCD camera and the software "ImageJ" was used to analyze the magnetic particles density over the gold surfaces. The processed chips were measured electrically as described later.

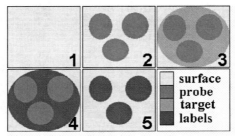

FIGURE 2. Schematic representation of the procedure used to perform probe spotting and biomolecular recognition on the surface of dummy gold substrates. 1) Cleaned gold substrate; 2) Discrete probe spots of 2.5 μL; 3) Biorecognition, the larger circle represents the target solution covering the probe spots; 4) Magnetic labeling with magnetic particles suspension; 5) Remaining magnetic particles attached to the target molecules after washing.

2.3.1. Oligonucleotides Hybridization Assays

A referenced protocol [19] based on TE-HCl buffer was adapted and used as a starting point to understand the salt effect on the final density of magnetically labeled hybridized DNA targets. Briefly, immobilization and blocking solutions of thiolated - ssDNA and –PEG, respectively, were prepared in TE-HCl (see formulation on section 2.2) with 1 M of K_2HPO_4, pH 7.4; the hybridization in solution of complementary ssDNA was performed in TE-HCl with 1 M NaCl, pH 7.0; the magnetic labeling and final washing with TRIS-HCl 100 mM, pH 8.0, with 1 M NaCl and 0.02% Tween20®. The absence of salt during the hybridization step and the gradual decrease in salt concentration on the biotin/streptavidin-based labeling step was evaluated.

Results are presented on the graph from Figure 3 where the percentage of magnetic labels density should not be seen as an absolute value, but as a relative measurement due to the small size of the used particles.

From Figure 3 the main findings are the following: i.) the absence of NaCl either in the hybridization or labeling step reduces the number of recognition events to almost nil; ii.) the gradual decrease of NaCl concentration from 1 M down to 0.1 M in the labeling step corresponds to a proportional decrease of the density of magnetic beads; iii.) while keeping constant the salt concentration and varying the DNA probe concentration on the immobilization step from 0.1 to 4 μM, a maximum signal was found for a DNA probe at 1 μM.

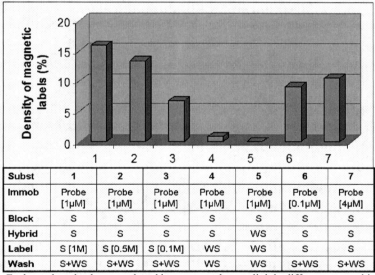

Subst	1	2	3	4	5	6	7
Immob	Probe [1µM]	Probe [1µM]	Probe [1µM]	Probe [1µM]	Probe [1µM]	Probe [0.1µM]	Probe [4µM]
Block	S	S	S	S	S	S	S
Hybrid	S	S	S	S	WS	S	S
Label	S [1M]	S [0.5M]	S [0.1M]	WS	WS	S	S
Wash	S+WS	S+WS	S+WS	WS	WS	S+WS	S+WS

FIGURE 3. Each numbered column on the table corresponds to a slightly different protocol in terms of salt concentration in the buffer solutions of TE-HCl used at ssDNA oligonucleotides hybridization, streptavidin/biotin labeling with 2 μm magnetic labels and washing steps. The nomenclature used in the table stands for: Probe [X μM] – DNA probe concentration; [S] – presence of salt, in the blocking step stands for 1M K_2HPO_4, in the hybridization, labeling and washing steps it stands for 1M NaCl; [WS] – absence of salt; S [1M], S [0.5], S [0.1] – NaCl concentrations, respectively; [S+WS] – sequential wash with buffer at 1M NaCl and without salt. Each bar in the graph corresponds to the density of magnetic particles on the gold surfaces submitted to the correspondent protocol in each column of the table.

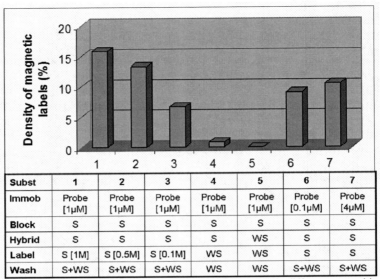

Subst	1	2	3	4	5	6	7
Immob	Probe [1µM]	Probe [1µM]	Probe [1µM]	Probe [1µM]	Probe [1µM]	Probe [0.1µM]	Probe [4µM]
Block	S	S	S	S	S	S	S
Hybrid	S	S	S	S	WS	S	S
Label	S [1M]	S [0.5M]	S [0.1M]	WS	WS	S	S
Wash	S+WS	S+WS	S+WS	WS	WS	S+WS	S+WS

FIGURE 4. Each numbered column on the table corresponds to a slightly different protocol in terms of salt concentration and type of buffer solutions used (TE or PB buffer). The numenclature used in the table stands for: S – TE-HCl buffer with 1M K_2HPO_4; WS (PB) – phosphate buffer without salt; S(TE) – TE-HCl buffer with 1M NaCl; S (PBS) – phosphate buffer saline 150mM NaCl. Each bar in the graph corresponds to the density of magnetic particles on the gold surfaces submitted to the corresponding protocol in each column of the table.

In Figure 4 another set of experiments on the salt effect is presented, this time we were replacing the TE-HCl buffer by phosphate buffer (PB) and the TE-NaCl (1 M NaCl) by PBS (0.15 M NaCl) in certain steps of the functionalization, but we were always keeping constant the immobilization buffer (TE-HCl, 1 M K_2HPO_4, pH 7.4). In this case, despite the loss of magnetic labels density (down to 80% less) when PBS or PB substituted TE-NaCl, the remaining signal of recognition events is still measurable. This does not hold when non-salty TE is used (Fig. 3).

Assuming a compromise between the density of recognition events and the harshness level of the chemistry on the chip, in following experiments the adopted protocol was the one represented on column 3 from the table on Figure 4.

The fully optimized protocol consists of the following steps: i.) thiol-modified DNA probes at 1 µM in TE-HCl are spotted on the N_2 stream dried chip surface and left to immobilize at 37°C; ii.) after 2 hours the remaining unbound DNA strands are washed away with the same buffer; iii.) a blocking step with SH-PEG 1mg/mL in TE-HCl, 1h at 37°C, is employed to minimize unspecific binding from both target molecules and magnetic labels; iv.) hybridization is carried out with biotinylated single stranded DNA target molecules at various concentrations on PB for 3 hours in a humid chamber at 42°C; v.) loose and/or not specifically bound biomolecules are removed by using a PB with 0.02% (v/v) Tween20®.

2.3.2. "Sandwich" Immuno-recognition Assays

For the preparation of immunosensitive surfaces, specific antibodies against common structure antigens (CSA) of *Salmonella sp.* whole cells were covalently immobilized on the gold surface through a random method. A heterobifunctional linker, the SPDP, containing a disulphide bond (-S-S-) was used to introduce thiol functional groups in the antibody molecule. In detail, an aqueous suspension of SPDP at 4 mg/mL was added on a 1:3 (v/v) ratio to the antibody solution at 0.5 mg/mL in phosphate buffer (20 mM), 150 mM NaCl, pH 7.4, and let to react under gentle agitation for 2 hours at room temperature (RT). After that, the antibody is purified using a centrifugal filter unit of 30 kDa molecular weight cut-off (Centricon®, Millipore), washed and resuspended at ~0.5 mg/mL in sodium acetate buffer 0.1M, 0.1M NaCl, pH 4.5. To the purified antibody the reducing agent dithiothreitol (DTT) is added to a final concentration of 50 mg/mL and let react under gentle agitation for 30 min at RT. After reduction the DTT is removed from the antibody using another Centricon, washed and resuspended at 0.5 mg/mL in TE-HCl buffer.

All the centrifugation steps are performed at 10,000 rpm in a refrigerated centrifuge at 21°C.

The antibody is spotted on the chip surface and let to immobilize for 2 h at 37°C in a humid chamber. After antibody immobilization the remaining free binding sites were inactivated using a blocking solution of SH-PEG 1 mg/mL, BSA at 10 mg/mL or a mixture of both SH-PEG and BSA on TE-HCl buffer that reacted for 1 h at 37°C. After this, a droplet of 50 μL concentrated *Salmonella* cells suspension in PBS, 0.02% Tween20®, was placed on top of the antibody-modified chip surface for 1 h at RT. The unbound species were washed away from the surface with PBS-Tween20®. Subsequently, a second polyclonal anti-*Salmonella* antibody, bearing a biotin modification, was spotted at 25-50 μg/mL in PB-Tween20® and left to react for 2 h at RT. At the end, a final rinse with PB-Tween20® is used to remove all unbound species from the sensing surface.

During the optimization of the surface biochemistry for protein chips two main aspects were considered and evaluated: i.) the cleaning procedure of the bare gold surfaces and ii.) the blocking solutions applied between probe immobilization and target recognition.

As a mean to choose an efficient cleaning procedure, before applying the previously described biorecognition protocol, biotinylated antibodies were immobilized to gold surfaces treated with three different cleaning methods. The cleaning methods were compared in terms of maximum achievable density of 2 μm sized streptavidin magnetic beads. Those values can be directly translated into the density of immobilized probe molecules. In Figure 5 the Piranha solution (30% H_2O_2/ 70% conc. H_2SO_4) is compared with concentrated HCl and Oxygen plasma dry cleaning procedures. A gold substrate cleaned by the standard wet-bench procedure, Microstrip-2001® (Fuji-Film, Portugal) at 80°C for 2 h rinsed with IPA and water, was used as a reference result. After the cleaning step, substrates were submitted to immobilization of biotinylated antibodies on the chip surface followed by magnetic labeling. The O_2 plasma achieved a significantly higher particles density on the spotted area while keeping the non-specific background nil. Both, non-thiolated

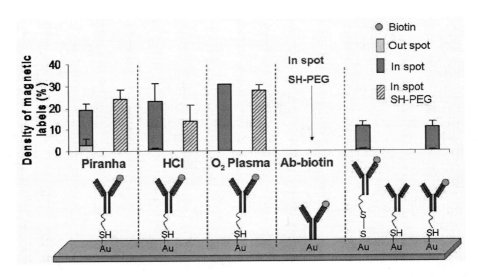

FIGURE 5. Comparison of three different cleaning methods efficiencies in terms of density of streptavidin modified magnetic labels recognizing immobilized biotinylated antibodies.

biotiniylated antibodies and thiolated non-biotinylated antibodies do not adsorb to the gold surface or facilitates magnetic beads adsorption, respectively. However, non-reduced SPDS-derivatized antibodies were found to bind to the spotted area, meaning that in some extent the disulphide linkage is reduced even in the absence of the DTT reducing agent as reported elsewhere [23]. Furthermore, at this level of functionalization, the blocking step do not seems to be crucial to avoid non-specific binding. Non-blocked (solid bars, Figure 5) and SH-PEG blocked (striped bars, Figure 5) antibody-derivatized surfaces originate very similar responses in terms of particles densities, both inside and outside the antibody spotted area.

Another set of experiments was carried out to visualize and quantify the *Salmonella* cells adhesion in the presence of SH-PEG and BSA blocking molecules as well as the adsorption (cross-reactivity) to a non-specific antibody. For this purpose probe spotting and biorecognition proceeded as described in Figure 2. The blocking agents were chosen attending that polymers such as PEG and BSA proteins are known as intrinsically inert towards the adsorption of proteins, cells and other biomolecules, thus providing a zero background starting surface and avoiding false positive signals.

Two different probe antibodies were immobilized on distinct chips. One of the probes was a specific antibody for *Salmonella* cells (positive control) and the other one a human IgG non-specific for *Salmonella* (negative control). Then, *Salmonella* cells were captured and magnetically labeled with 2 μm streptavidin modified magnetic beads by a so-called "sandwich" type immunoassay. At the end, substrates were observed under an optical microscope at a 40x magnification and pictures were recorded (Fig. 6). In fact, using this blocking solution the non-specific adhesion out-side the spotted area was greatly minimized. Another important aspect is the target cells concentration on the measurable sample. Two different sample concentrations (concentrated sample and 10,000x diluted) were compared in terms of non-specific

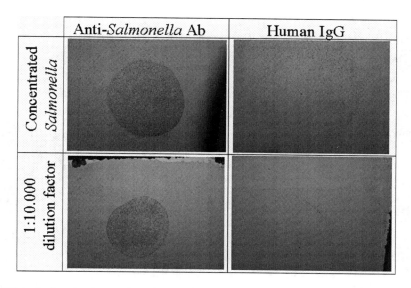

	Anti-*Salmonella* Ab	Human IgG
Concentrated *Salmonella*		
1:10.000 dilution factor		

FIGURE 6. Surface density of 3 μm particles labeling Salmonella cells immobilized on gold surfaces. First column shows spots of specific antibodies anti-Salmonella (positive control); second column shows spots of non-specific human IgG (negative control). Substrates corresponding to images on the first row were reacted with a concentrated *Salmonella* suspension, on the second row the *Salmonella* suspension was 10,000x less concentrated. Surfaces were blocked wit SH-PEG/BSA and immobilized *Salmonella* cells magnetically labeled by biotinylated antibodies bearing a streptavidin magnetic label.

adsorption and cross-reactivity to the negative control spotted area. The concentrated sample was prepared resuspending a loop-full of *Salmonella* cells in a 1 mL of PBS-Tween20®. In Figure 6 is shown that both non-specific adsorption and cross-reactivity were proportionally reduced with sample concentration.

Although, such level of contamination is not a common scenario in water for human consumption and false negatives are much more critical than false positives, some more effort should be done at the blocking and washing steps level. Additionally, using a careful selection of monoclonal antibodies instead of polyclonal, among other advantages, should significantly reduce cross-reactivity.

2.4. Electronic Read-out

A pre-functionalized chip was mounted on a breadboard and coaxial cables were used to make electrical connections to the associated hardware, namely to power sources and to a general purpose interface bus (GPIB)-controlled lock-in amplifier (Fig. 7). Measurements were made using an in-plane transverse external AC excitation field of 13.5 Oe rms (31 Hz) in combination with a DC bias field of 10 - 30 Oe to magnetize the superparamagnetic particles. The optimum DC field is determined by the analysis of the SV transfer curve (Fig. 8) and corresponds to the value where the sensor starts entering in its linear range [22]. The sensor was biased with a 1 mA current. The AC signal coming from the sensor was acquired by a lock-in technique.

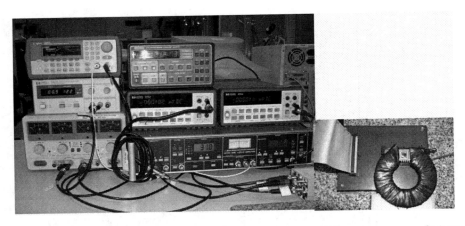

FIGURE 7. Experimental setup comprising a function generator, two DC current power supply (one to the sensor and other to the coil), a voltimeter, an amperimeter, a lock-in amplifier, a multiplexer and an horse-shoe electromagnet.

The magnetic labeling of target molecules can be done either before or after biomolecular recognition. In these particular experiments the labeling was done after the biorecognition event took place.

Immediately before using the magnetic particles, its original preservative solution was removed and particles were carefully washed and resuspended in PB/Tween20® buffer solution. Two different sizes of streptavidin-modified magnetic labels were used. 250 nm particles and micron-sized, 2-3 µm, beads were used in the labeling of biotinylated DNA molecules and biotinylated anti-*Salmonella* antibodies in cell detection, respectively.

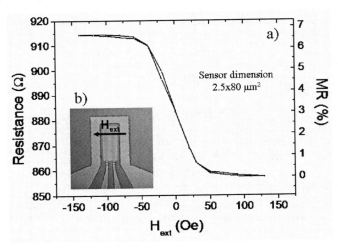

FIGURE 8. a) Spin valve transfer curve measured applying a 1mA current bias. The spin-valve active area dimension is 2.5 x 80 µm^2 (u-shaped); inset b) Top view of the spin-valve with indication of the direction of the external field applied to the sensor during the measurements.

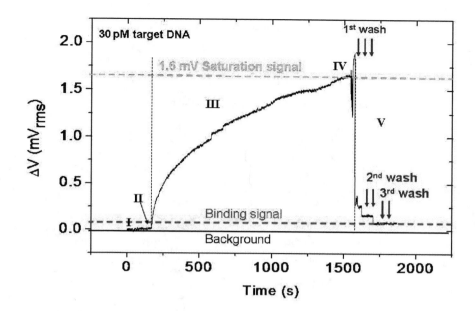

FIGURE 9. Real-time data taken from a single spin-valve sensor on the detection of Nanomag®-D 250 nm magnetic particles involved in the labeling of a complementary target hybridization event. The main phases in the signal acquisition procedure are indentified by: I) Baseline stabilization; II) Magnetic particles addition; III) Increasing signal correspondent to particles settling down over the sensor; IV) Saturation signal; V) Washing steps.

Figure 9 shows a typical output of a single sensor. Different phases of the signal acquisition correspond to the following procedure: Phase I.) Initially the baseline signal is let to stabilize with a 50 μL PB droplet over the chip; Phase II.) The droplet is replaced by a 50 μL droplet of specifically modified magnetic labels 5x diluted from the stock solution at 10mg/mL; Phase III.) Particles are left to settle down and interact with immobilized targets for 15 minutes, corresponding to an increase in the differential voltage signal; Phase IV.) The saturation signal is attained when the signal stabilizes and is recorded for *a posteriori* normalization of binding signals between different sensors and chips; Phase V.) Finally, the sensor is double or triple washed with PB/Tween20® to remove unspecific, weakly bound particles and to confirm the stability of the binding signal.

2.5. *E. Coli* 16S Oligo Sequences Hybridization Detection

Two groups of experiments with results on the magnetic detection of hybridization events using spin-valve based biochips were performed. A first group was performed using two different target DNA sequences, complementary and non-complementary, at a constant concentration of 1 μM. This experiment serves to assess the reproducibility (measuring the same target with different chips), background signal and non-specific response of the standard biochip platform in use. As depicted in Figure 10 the amplitude of the positive magnetic response (hybridization between fully

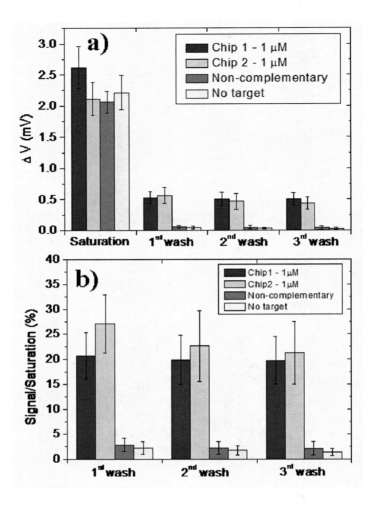

FIGURE 10. Voltage data from magnetic labels detection on single stranded oligonucleotides hybridization experiments with 1 μM target molecules. The signals are average values from the sequential read-out of 16 sensors. a) Signals in voltage (mV) scale as acquired from the reading set-up; b) Signals normalized to the correspondent saturation signal of the sensors.

complementary strands) reaches about 500 μV out of ~2500 μV saturation signal. Comparing this signal to the 40 and 30 μV average signals resulting from a negative control (where a 30% complementarily target strand is used) and a blank (where no target is present), respectively, a difference of more than 20% in signal amplitude is found. Also signals from different chips measuring the same target concentration (1 μM) were compared before and after normalization to the saturation signal. Variations in the average signal of different chips of 6.5, 3 and 1.5% for the first, second and third washing steps, respectively, were recorded, indicating a significant reduction of non-specific binding along the washing procedure.

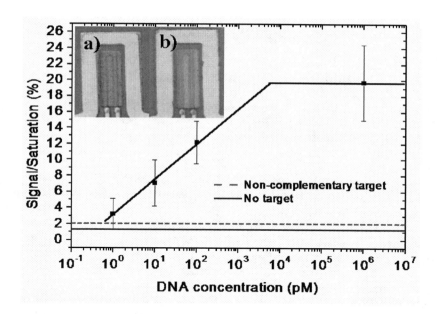

FIGURE 11. Biological detection limit for single stranded DNA sequences encoding for the genomic region of the 16S ribosomal sub-unit of *E. coli*. The straight lines connecting the experimental data points are simply a guide to the eye. *Inset:* Optical microscope pictures at 800x magnification, from individual sensors corresponding to a) 1 μM target and b) no target assay.

A second group of hybridization experiments was performed to identify the biological detection limit of the standard platform operating as an oligonucleotide hybridization detection system, using 250 nm in diameter magnetic labels. Solutions of target DNA were prepared at exact concentration diluted from a stock solution of 100 μM.

By analyzing the graph from Figure 11 a linear sensor response was achieved for non-saturating concentrations and 1 μM was found to be a saturating concentration. A picomolar target solution is on the edge of the detection limit and its associated standard deviation crosses already the boundary to the non-complementary signal. The fitting curve shows that a discrimination of concentrations ranging from 1 pM to 10 nM can be accomplished. Further measurements should be performed in the intermediate concentrations, especially between 10^2 and 10^4 pM, to precisely determine the saturating concentration.

The standard deviations associated to these signals are relatively high, situated in between 2 and 5% for 1 pM and 1 μM, respectively. A possible reason for this is the manual washing procedure that was used. In order to have a more reproducible washing procedure, this manual washing will be replaced in the near future by a controllable microfluidic washing system.

3. MTJ-BASED BIOCHIP

3.1. MTJ Sensors and Chip Microfabrication

This biochip consists of 32 MTJ probe sites, with each MTJ surrounded by a U-shaped current line. Magnetic Tunnel Junctions (MTJ) sensors were deposited onto Corning Glass 7059 substrates with the following structure: Ta 30 Å/ Ru 150 Å/ $Mn_{76}Ir_{24}$ 150 Å/ $Co_{56}Fe_{44}$ 40 Å/ Ru 8 Å/ $(Co_{70}Fe_{30})_{80}B_{20}$ 50 Å/ Al 6 Å (+oxidation)/ $(Co_{70}Fe_{30})_{80}B_{20}$ 50 Å/ Ru 100 Å/ Ta 30 Å. These layers were deposited by ion beam in a Nordiko 3000 system and passivated with a 150 Å thick $Ti_{10}W_{90}(N)$ antireflective coating. The easy axis of the magnetic layers was defined during deposition by a 40 Oe magnetic field (parallel anisotropies, linear response obtained from shape anisotropy) and the oxidation of the barrier was obtained by a remote Ar/O_2 plasma applied for 15 seconds during the Al deposition. After defining 32 MTJ areas with 2 x 15 μm^2 (sensing area) by ion milling, a 500 Å thick Al_2O_3 layer was sputtered in order to isolate laterally the structures. The top contact leads of the MTJs as well as the U-shaped current lines surrounding the sensors (also present in the spin valve biochip) were defined by lift-off of a $AlSi_1Cu_{0.5}$ 2000 Å/ $Ti_{10}W_{90}(N)$ 150 Å layer deposited by magnetron sputtering (Nordiko 7000 system). In order to prevent corrosion, an Al_2O_3 2000 Å / SiO_2 1000 Å stack was deposited by RF sputtering over the whole chip leaving open contacts for the electrical biasing of the MTJ. For further detail, a cross section of the biochip sensing site is shown in Figure 12. The sample was annealed in vacuum at 280°C under an 1 Tesla magnetic field parallel to the MTJ pinned layer magnetization direction to improve the exchange bias. Finally, as done with the spin valve biochip, the MTJ biochip was wire-bonded on a chip-carrier for further experiments.

A typical transfer curve for these sensors is shown in Figure 13, with an MR of 43%, a resistance (in the parallel state) of 319 Ω and a sensitivity of 0.23%/Oe.

The MTJ were used to detect 130 nm particles. The original particle preservative solution was replaced by a PB solution. The particles were then diluted on the ratio of 1:10 ($2x10^{11}$ particles/mL) and 1:100 ($2x10^{10}$ particles/mL) in PB.

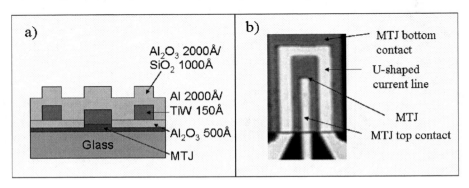

FIGURE 12. Crossectional (a) and top (b) view of the sensing site of a MTJ.

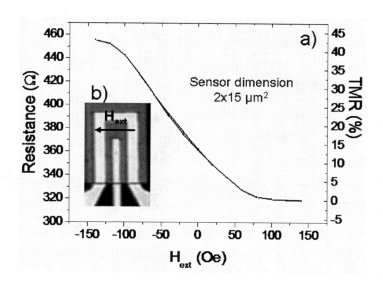

FIGURE 13. a) Magnetic tunnel junction transfer curve as measured by applying a 200 μA current bias. The spin valve active area dimension is 2 x 15 μm². b) Top view of the MTJ with indication of the direction of the external field (H_{ext}) applied to the sensor during the measurements.

3.2. MTJ Particles Detection

The particle detection was achieved by using a handheld electronic platform. The platform (Fig. 14) integrates all the electronic circuitry required for generating the sensor bias current and the DC+AC in-plane external magnetic field as well as the electronics needed for signal acquisition and processing. High-level system control and data analysis are remotely performed through a laptop via a serial communication channel [12]. In this specific experiment, the MTJ was biased with a 200 μA current and a 25 Oe + 15 Oe_{rms} (375 Hz) DC+AC magnetic field which was applied in the short direction of the sensor to magnetize the particles. The resulting signal was amplified and converted to digital configuration allowing the usage of a digital narrow pass-band filter with a bandwidth of 0.25 Hz. Before putting particles, the signal was acquired and averaged during 10 minutes to define the baseline signal of the sensor. The different concentrations (ranging from $2x10^{12}$ to $2x10^{10}$ particles/mL) of 130 nm magnetic particles were then put on the chip surface. After waiting 20 minutes for the particles to settle down, the signal was again acquired and averaged during 10 minutes. The difference between this signal and the baseline signal is the saturation signal, *i.e.* the maximum signal that the sensor can measure for the used particle concentration.

External electromagnet

Communication and electromagnet driving board

Acquisition board

Biochip

FIGURE 14. Portable electronic platform comprising a socket to put in the biochip, an acquisition board, a communication and electromagnet driving board, and an external electromagnet.

Figure 15 shows that the saturation signal is decreasing with the concentration of particles. The noise of the platform is limiting the detection to a concentration of 2×10^{10} particles/mL. An increase in particle sensitivity is expected, if the sensor current bias is increased to 900 µA. This value corresponds to the current for which the MTJ show a maximum signal to noise ratio (Fig. 16). Above this value, the signal to noise ratio of the MTJ is decreasing due to the decrease of its magnetoresistance with the increasing of the current. Improvements in the platform noise are also under investigation.

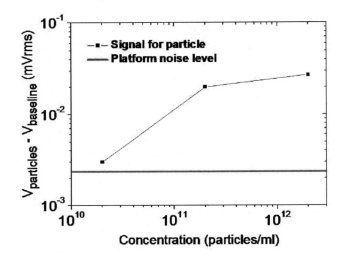

FIGURE 15. Magnetic tunnel junction voltage variation as a function of the concentration of 130 nm magnetic particles and the noise level of the measuring platform.

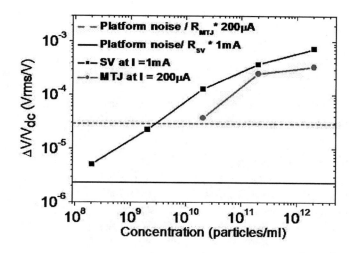

FIGURE 16. Detection of 130 nm particles at dilutions 1:1, 1:10, 1:100, 1:1000 and 1:10000 by MTJ (circles) and SV (squares) sensors using the electronic platform from Figure 14. In order to be comparable, the signals were normalized to the DC voltage of the sensor. The noise was also normalized and indicates the detection limit of the system.

3.3. Spin-valve and MTJ Sensor Comparison

In this section a comparison between the spin-valve and MTJ sensor particle sensitivity is made. For this purpose, the integrated platform used in the previous section was used for the detection of several concentrations of 130 nm particle solutions (ranging from $2x10^{12}$ to $2x10^{10}$ particles/mL) and for each sensors signal-to-noise ratio (SNR) characterization.

The SNR characterization was made using an AC in-plane magnetic field of 2.4 Oe_{rms} (375 Hz). The signal was digitally filtered using a narrow pass-band filter with a bandwidth of 0.25 Hz. The noise (dominated by the electronic platform) and signal levels were averaged for a 10 minutes acquisition window.

Figure 17 compares the SNR obtained for spin-valves and MTJs. As observed, for the same current the MTJ SNR is always higher than the spin-valve SNR. The spin-valve SNR is increasing linearly with the current while the MTJ SNR shows a maximum value at 900 µA. The limitation for biochips using spin valves is related to the heat generation due to the Joule effect and electrical failure through the fluid (breakdown occurs for bias voltages of 1-2 V).

For particle detection, the platform was configured to deliver a bias current of 200 µA and 1 mA for MTJ and SV sensor, respectively. The generated AC and DC magnetic fields were 15 Oe_{rms} and 25 Oe, respectively. The particles were measured at dilutions of 1:10 ($2x10^{11}$ particles/mL), 1:100 ($2x10^{10}$ particles/mL), 1:1000 ($2x10^9$ particles/mL) and 1:10000 ($2x10^8$ particles/mL) in PB and the signal was processed in the way described in section 3.2. The saturation signal was normalized to the sensor DC voltage allowing the comparison between MTJ and the SV signals. As shown on

Figure 16, the SV detection limit corresponds to a dilution of 1:10000 (2 x 10^8 particles/mL), while for the MTJ this detection limit was for a dilution of 1:100 (2 x 10^{10} particles/mL). This is due to the fact that a lower current bias was used to make the measurements with the MTJ. If this current bias is increased to 900 μA (maximum SNR), a corresponding increase of the signal obtained by the presence of magnetic particles is expected. Furthermore, for the same particle concentration the SV chip shows higher particle sensitivity than the MTJ chip. However, looking at the SNR of these two sensors, a higher signal in the MTJ chip was expect. This discrepancy arises from the extra spacing between magnetic labels and the active free layers in MTJs when compared with SVs. This extra spacing appears because the MTJ fabrication requires a metal contact (2000 Å Al) between the sensing layer and the oxide (see section 3.1) while this do not appear in the spin valves structure (see section 2.2). A possible improvement of the MTJ particle sensitivity was pointed out in reference [24], where gold electrodes replaced the aluminum contact covered with the passivation layer. This seems to be a suitable solution for our system knowing that the chemistry in use requires a gold surface. However, tests are still required to check possible sensor corrosion issues with this geometry.

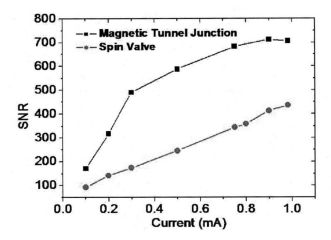

FIGURE 17. Magnetic tunnel junction and spin-valve signal-to-noise ratio as a function of the applied bias current as measured with the electronic platform.

4. SYSTEM INTEGRATION

In miniaturized, portable, easy-to-use and automated devices, two important issues are here identified to be addressed: i.) the scalability of the detection features, seeking high throughput applications and ii.) the microfluidic system for sample preparation, separation, concentration and transportation.

The biochips reported in the last years were designed with few sensing elements (25 in the INESC-MN cystic fibrosis biochip [15], 64 sensors in the BARC biochip [5] and 206 sensors in the biochip developed at the University of Bielefeld [6]). Although the number of sensors used in these biochips may be enough for several applications, a bioplatform should present a large dynamic range in terms of discrete sensing elements in order to be versatile enough to respond to as many as possible different applications. It may be possible to inquire a water sample searching for few different pathogenic microorganisms or to simultaneously screen hundreds of mRNA or cDNA sequences for genetic mapping using exactly the same bioplatform. In order to make this possible the sensing elements should be miniaturized and highly repeated in a dense matrix, where each element is individually addressed and read-out in almost real-time. Taking this into account, matrix based biochips were recently proposed [13, 25].

Additionally, a microfluidic system is a key aspect to the successful development of the final apparatus. As a mean to facilitate and at the same time make sample transportation a more controllable and reproducible task, two different, but complementary, issues are being addressed. Presently there are two independent microfluidic units under development and microfabrication: i.) a microbial cell concentrator that presently promises the ability to concentrate 100 mL sample volumes down to 100 μL and ii.) a magnetic microfluidic separator for cell sorting and concentration. The fabrication of the cell concentrator and separator comprises two stages: the spin-valve sensing unit and electromagnetic transport fabrication and the microfluidic system itself in polydimethylsiloxane (PDMS) fabrication.

4.1. MTJ/Diode Array

The approach followed at INESC-MN to accomplish the production of a matrix of 16x16 sensing sites uses a basic cell element composed by an amorphous silicon thin film diode connected in series with a MTJ. The proposed platform not only integrates the array of magnetoresistive sensors, but also provides all the electronic circuitry for addressing and read-out each magnetic transducer.

Figure 18 shows a detailed cross section of the biochip at a sensing site for two different types of diodes. The MTJ was chosen because of its flexibility in controlling the resistance and its high sensitivity. A diode switching element was used rather than a transistor in order to avoid additional control lines, simplifying the process.

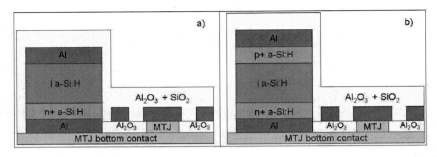

FIGURE 18. a) Film stack of a matrix element composed by a MTJ and an a-Si:H Schottky diode.
b) Film stack of a matrix element composed by a MTJ and an a-Si:H PIN diode.

With the technology available at INESC-MN, two different kinds of diodes can be fabricated i.e. PIN and Schottky diodes. Table 1 shows the maximum current, the ON/OFF ratio and the process yield of each diode. A very important quantity for the use of a diode as switching element on a matrix is its process yield. If one of the diodes in the matrix is not working (*i.e.* the diode does not rectify), leak current will pass in this element while addressing any other element in the matrix. The yield of the diode process has to be 100%. Taking this into account, the PIN diodes are the best option for matrix applications. The PIN diodes are also more robust (can handle higher currents) and show a higher ON-OFF ratio enabling a larger number of elements in the matrix when compared with the Schottky diode. However, the Schottky diodes present lower noise levels than the PIN diodes (Figure 19). As discussed in [26] this fact may not be a problem, because for frequencies above 100 Hz a single 250 or 130 nm particle (nanomag-D, Micromod) can be detected.

TABLE 1. PIN and Schottky diode yield, ON/OFF ratio and maximum current comparison.

	Max. Current	On/Off Ratio	Yield
Schottky	0.2 mA	10^6	68.8%
PIN	1 mA	10^8	98.4%

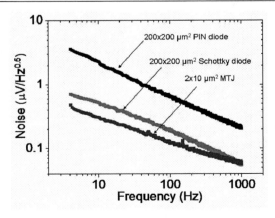

FIGURE 19. Noise spectra at a 15 Oe field of a 30 kΩ MTJ, with a Hooge constant of 5×10^{-9} μm^2 and an area of 2 x 10 μm^2, and of a 200 x 200 μm^2 PIN and Schottky diode.

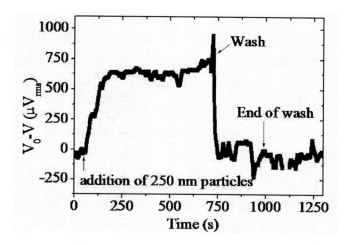

FIGURE 20. Voltage variation due to the insertion of 10 μL of 1:10 stock dilution of 250 nm particles.

Finally, 250 nm particles detection was performed with one matrix element. These particles were prepared in the same way as previously explained and only a dilution of 1:10 was tested. A 6 Oe + 15 Oe$_{rms}$ (30Hz) DC+AC magnetic field was applied to magnetize the particles and an external lock-in technique was used to readout the data shown in Figure 20. A saturation signal of 650 μV$_{rms}$ was obtained. The noise is still high in this experiments but this aspect may be improved by under-taking the experiment in the electronic platform described in section 3.2.

4.2. Macro/Micro-fluidic Concentrator

The concentrator design is based on a microfluidic system which consists of a long channel with concentration net channels at its end. The length of the channel is taken to be 1.6 cm before the beginning of the concentrating net channels, to assure the focusing of the cells at the centre of the main channel. The concentration net channel is approximately 0.02 cm long. The overall design is displayed in Figure 21. The system comprises one inlet reservoir and three outlet reservoirs; in the middle the concentration outlet and on the sides the discharge outlets. The small structures before the outlets are the net channels mainly responsible for the removal of unwanted fluid volumes.

To test if the fluidic system has the ability to reduce the volume of the inlet sample, preliminary tests with 1 mL samples of PB were done. In this test, a 1mL sample was inputted in the microfluidic system using a syringe pump and the outlets measured. Since only few μL will be flowing out, a caliper was used to measure the length of the concentrated fluid column in the tubage. Knowing that the tubage has a diameter of 812 μm, the output volume can be estimated. The maximum and minimum measured volumes were 2.12 μL (4.1 mm measured in the fluid column) and 1.07 μL (2.1 mm measured in the fluid column), respectively. After several measurements, a 1.5 μL average concentrated volume was determined. From this the calculated concentration

171

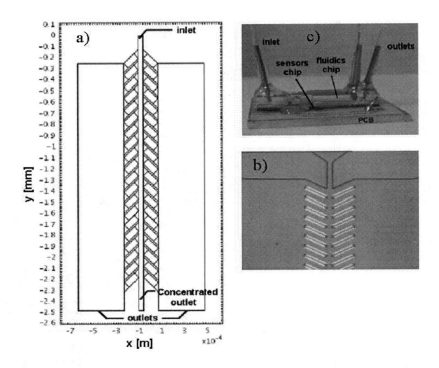

FIGURE 21. a) Design of the fluidic system of the concentrator. b) PDMS patterned with the concentrating net channels. c) Magnetic cell concentrator assembled to a PCB backbone and with the connector for macroscopic fluidic connections.

factor was in the order of 700x, which is in fact the desired order of concentration. Further tests involving the chip electromagnetic transportation lines need to be done.

4.3. Cell Magnetic Separation

An integrated microfluidic system based on magnetophoresis is being developed using *in-situ* magnetic fields generated by current lines (a magnetic field gradient (∇H) of 9×10^8 A/m^2 and a gradient (H) of 2.3kA/m is created by applying 40 mA). The chip is composed by a separation stage and a counter stage. In the separation stage, magnetically labelled cells will be separated from other cells and dispersed beads (moving from one channel to the other). Then, the separated cells are counted by a spin-valve sensor in the counter stage. Spin-valve sensors are designed in such a way that, for a bias current of 5 mA, a signal of 320 μV/cell is measured when a cell passes over it [27-29]. The microfluidic system has been fabricated in PDMS and consists of two channels (40 μm wide, 14 μm thick and 20 μm apart) joined over a gap of 2 mm, in which the fluid is laminar [30]. A 200 μm/s velocity is required in order to have an effective separation (Fig. 22). Both parts of the system can be irreversibly bonded to each other. This system is at the moment being tested for the isolation of human hematopoietic stem/progenitor cells from umbilical cord blood samples.

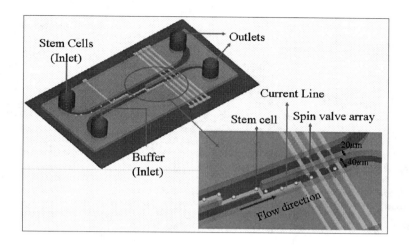

FIGURE 22. "H-type" fluidic platform allowing magnetically labeled cells to be separated from the left to the right channel due to the magnetic field created from the current line. Each channel has spin-valves at the end to count the cells.

5. CONCLUSIONS

Surface biochemistry was intensively studied and optimized, resulting in controllable and stable DNA arrays of fully active bioreceptors that result in a detection limit for the standard platform on the picomolar range. On the other hand, protein biochips remain under optimization aiming a less noisy background and more reproducible signal. So far we have partially accomplished the main tasks for the development of the bioelectronic device. The chip integration in an electronic platform with a microfluidic system is being actively pursued seeking for an apparatus that intend to be highly scalable, fully automated and portable, integrating the lab-on-a-chip concept.

ACKNOWLEDGMENTS

The authors acknowledge the financial support of the European projects EC FP6 under the contracts NMP4-CT-2005-016833 (SNIP2CHIP) and NMP4-CT-2005-017210 (Biomagsens). V.C. Martins, F.A. Cardoso and J. Loureiro thank FCT, Portugal for SFRH/BD/13725/2003, SFRH/BD/23756/2005 and SFRH/BD/30056/2006 doctoral grants, respectively.

REFERENCES

1. P.P. Freitas, R. Ferreira, S. Cardoso and F.A. Cardoso, "Magnetoresistive sensors", *J. Phys.: Condens. Matter* **19**, 1-21 (2007).
2. D.L. Graham, H.A. Ferreira and P.P. Freitas, "Magnetoresistive-based biosensors and biochips", *Trends Biotechnol.* **22**, 455-462 (2004).

3. G. Li, S. Sun, R.J. Wilson, R.L. White, N. Pourmand and S.W. Wang, "Spin valve sensors for ultrasensitive detection of superparamagnetic nanoparticles for biological applications", *Sens. Actuat. A Phys.* **126**, 98-106 (2006).

4. M.M. Miller, P.E. Sheehan, R.L. Edelstein, C.R. Tamanaha, L. Zhong, S. Bounnak, L.J. Whitman and R.J. Colton, "A DNA array sensor utilizing magnetic microbeads and magnetoelectronics detection", *J. Magn. Magn. Mater.* **225**, 138-144 (2001).

5. J.C. Rife, M.M. Miller, P.E. Sheehan, C.R. Tamanha, M. Tondra and L.J. Whitman, "Design and performance of GMR sensors for the detection of magnetic microbeads in biosensors", *Sens. Actuators A* **107**, 209-218 (2003).

6. J. Schotter, P.B. Kamp, A. Becker, A. Pühler, G. Reiss and H. Brückl, "Comparison of a prototype magnetoresistive biosensor to standard fluorescent DNA detection", *Biosens. Bioelect.* **19**, 1149-1156 (2004).

7. J. Diao, D. Ren, J.R. Engestron and K.H. Lee, "A surface modification strategy on silicon nitride for developing biosensors", *Anal. Biochem.* **343**, 322-328 (2005).

8. L. Lagae, R. Wirix-Speetjens, C.-X. Liu, G. Borghs, S. Harvey, P. Galvin, H.A. Ferreira, D.L. Graham, P.P. Freitas, L.A. Clark and M.D. Amaral, "Magnetic biosensors for genetic screening of cystic fibrosis", *IEE Proc.-Circuits Devices Syst.* **152**, 393-400 (2005).

9. M. Megens and M. Prins, "Magnetic biochips: a new option for sensitive diagnostics", *J. Magn. Magn. Mater.* **293**, 702-708 (2005).

10. M.-L. Chan, G.M. Jaramillo and D.A. Horsley, "Scanning magnetic junction sensor for the detection of magnetically labeled DNA microarray", *IEEE CNF Transducers 2007*, 95-98 (2007).

11. H.A. Ferreira, F.A. Cardoso, R. Ferreira, S. Cardoso and P.P. Freitas, "Magnetoresistive DNA chips based on ac field focusing of magnetic labels", *J. Appl. Phys.* **99**, 08P105 (2006).

12. M. Piedade, L.A. Sousa, T.M. Almeida, J. Germano, B.A. Costa, J.M. Lemos, P.P. Freitas, H.A. Ferreira and F.A. Cardoso, "A new hand-held microsystem architecture for biological analysis", *IEEE Trans. Circ. Syst. I* **53**, 2384-2395 (2006).

13. F.A. Cardoso, H.A. Ferreira, J.P. Conde, V. Chu, P.P. Freitas, D. Vidal, J. Germano, L. Sousa, M.S. Piedade, B.A. Costa and J.M. Lemos, "Diode/magnetic tunnel junction cell for fully scalable matrix-based biochip", *J. Appl. Phys.* **99**, 08B307 (2006).

14. F.A. Cardoso, J. Germano, R. Ferreira, S. Cardoso, V.C. Martins, P.P. Freitas, M.S. Piedade and L. Sousa, "Detection of 130 nm magnetic particles by a portable electronic platform using spin valves and magnetic tunnel junctions sensors", *J. Appl. Phys.* **103**, 07A310 (2008).

15. H.A. Ferreira, N. Feliciano, D.L. Graham, L.A. Clarck, M.D. Amaral and P.P. Freitas, "Rapid DNA hybridization based on ac field focusing of magnetically labeled target DNA", *Appl. Phys. Letts.* **87**, 013901 (2005).

16. D.L. Graham, H.A. Ferreira, N. Feliciano, P.P. Freitas, L.A. Clarck and M.D. Amaral, "Magnetic field-assisted DNA hybridisation and simultaneous detection using micron-sized spin-valve sensors and magnetic nanoparticles", *Sens. Actuat. B Chem.* **107**, 936-944 (2005).

17. H.A. Ferreira, F.A. Cardoso, R. Ferreira, S. Cardoso and P.P. Freitas, "Magnetoresistive DNA chips based on ac field focusing and magnetic labels", *J. Appl. Phys.* **99**, 08P105 (2006).

18. [regulated by Decreto-Lei nº 236/98 de 1 de Agosto, DGA, 2000]

19. T.M. Herne and M.J. Tarlov, "Characterization of DNA probes immobilized on gold surfaces", *J. Am. Chem. Soc.* **119**, 8916-8920 (1997).

20. I. Park and N. Kim, "Thiolated salmonella antibody immobilization onto the gold surface of piezoelectric quartz crystal", *Biosens. Bioelectron.* **12**, 1091–1097 (1998).

21. A.W. Peterson, R.J. Heaton and R.M. Geargiadis, "The effect of surface probe density on DNA hybridization", *Nuc. Ac. Res.* **29**, 5163-5168 (2001).

22. H.A. Ferreira, N. Feliciano, D.L. Graham and P.P. Freitas, "Effect os spin-valve sensor magnetostatic fields on nanobead detection for biochip applications", *J. Appl. Phys.* **97**, 10Q904 (2005).

23. C. Vericat, M.E. Vela, G.A. Benitez, J.A.M. Gago, X. Torrelles and R.C. Salvarezza, "Surface characterization of sulfur and alkanethiol self-assembled monolayers on Au(111)", *J. Phys.: Condens. Matter* **18**, R867-R900 (2006).

24. S.X. Wang, S.-Y. Bae, G. Li, S. Sun, R.L. White, J.T. Kemp and C.D. Webb, "Towards a magnetic microarray for sensitive diagnostics", *J. Mag. Mag. Mater.* **293**, 731-736 (2005).

25. H. Shu-Jen, X. Liang, Y. Heng, R.J. Wilson, R.L. White, N. Pourmand and S.X. Wang, "CMOS integrated DNA microassay based on GMR sensors", *Elec. Dev. Meet. IEDM' 06*, 1-4 (2006).
26. F.A. Cardoso, R. Ferreira, S. Cardoso, J.P. Conde, V. Chu, P.P. Freitas, J. Germano, T. Almeida, L. Sousa and M.S. Piedade, "Noise characteristics and particle detection limits in diode+MTJ matrix elements for biochip applications", *IEEE Trans. Magn.* **43**, 2403-2405 (2007).
27. H.A. Ferreira, D.L. Graham, P. Parracho, V. Soares and P.P. Freitas, "Flow Velocity Measurement in Microchannels Using Magnetoresistive Chips", *IEEE Trans. Magn.* **40**, 2652-2654 (2004).
28. B.-H. Jo, L.M. Van Lerberghe, K.M. Motsegood and D.J. Beebe, "Three-Dimensional Micro-Channel Fabrication in Polydimethylsiloxane (PDMS) Elastomer", *(IEEE/ASME) J. MEMS* **9**, 76-81 (2000).

Moment Selective Digital Detection of Single Magnetic Beads for Multiplexed Bioassays

J. Llandro[a], T.J. Hayward[a], J.A.C. Bland[a], D. Morecroft[b], F.J. Castaño[c], I.A. Colin[c] and C.A. Ross[c]

[a]*Thin Film Magnetism & Materials Group, University of Cambridge, Cavendish Laboratory, JJ Thomson Avenue, Cambridge, CB3 0HE, UK.*
[b]*Research Laboratory of Electronics, Massachusetts Institute of Technology, 77 Massachusetts Avenue, Cambridge, MA 02139, USA.*
[c]*Department of Materials Science and Engineering, Massachusetts Institute of Technology, 77 Massachusetts Avenue, Cambridge, MA 02139, USA.*

Abstract. Research into lab-on-a-chip multiplexed bioassays has focused on libraries of biochemical probes, indexed by optically encoded micron-sized labels. However, few current methods have reconciled large multiplexing capability with a rapid detection system amenable to miniaturization. Magnetic identification of labels provides a strong candidate solution to this problem, yet no proposed single-label magnetic detection system can both read and encode magnetic labels. We present a magnetic multiplexed assay in lab-on-a-chip format which identifies target biomolecules from the hybridization results by reading encoded magnetic beads. We show that a microfabricated magnetoresistive ring-shaped sensor can read the magnetic moments of individual commercially available paramagnetic beads using an active digital technique. This work provides proof of principle for a new approach to magnetic labeling of biomolecules for high-throughput bioassays.

Keywords: Magnetic labeling, Spin-valve ring, Magnetic beads, Bioassay, Digital detection.
PACS: 75.47.De, 75.75.+a, 85.75.Ss, 87.85.Rs

INTRODUCTION

At present, research into realizing high-throughput multiplexed bioassays has concentrated on the development of microfabricated labels, which identify via a characteristic optical signature the single species of biomolecular probe (usually a short DNA sequence or protein) to which they are specifically bound. Optical encoding methods reported include spectrometric [1-5] (fluorophores, quantum dots and Raman tags), image-based or graphical [6, 7] and diffractive [8] identification. Fluorescent labeling in particular has become the current dominant technology in the field of molecular identification. However, strategies to minimize autofluorescent background [9] impose constraints on the portability of the detection system and sample preparation (due to the requirement for transparent substrates etc.). Spectral overlap limits multiplexing and errors are introduced by interference between signatures used to verify not only functionalization of the labels by the probes, but also hybridization between the probes and the targets, and finally to identify the probe (and thus the target). Graphical methods solve this problem by using fluorescence only for

CP1025, *Biomagnetism and Magnetic Biosystems Based on Molecular Recognition Processes*
edited by J. A. C. Bland and A. Ionescu

hybridization confirmation [7], but still require imaging of the label with an expensive and non-miniaturizable high-resolution CCD, with the associated disadvantages of non-real-time image analysis and data storage known from microarray methods.

The ideal combination of sensor and label would independently quantify the ability of probe molecules to bind both to the labels and targets, and also identify the targets from the hybridization results in a fully automated way. Magnetic beads (typically formed from iron oxide nanoparticles dispersed in a polymer or silica matrix) are a natural choice for such labels, due to their stability with respect to time, temperature, and reagent chemistry [10] and the lack of magnetic background generally found in biological samples. Conventional magnetic bead detection is commonly performed by placing a bead over a magnetoresistive sensor and applying a magnetic field H to generate a dipole moment m in the bead [11]. Because the magnetization aligns with the applied field, the stray field thus generated opposes and partially cancels the applied field below the bead over an area similar in size to its own cross-section. A simple model [12] for a uniformly magnetized paramagnetic sphere of radius a and susceptibility χ gives the fraction cancelled from the applied field as:

$$\frac{\Delta H}{H} = -0.38 \frac{\chi}{\chi + 3} \left(\frac{a}{z}\right)^3, \tag{1}$$

where z is the separation between the sensor and the centre of the bead and the prefactor [13] expresses the effect of averaging over the projected area. Despite intensive research done on magnetic biosensing, magnetic bead-based labeling has not been integrated successfully into multiplexed screening to date. Detection of the fields from such beads has relied on analogue sensors, the magnetoresistance (MR) of which is often carefully constrained to change linearly with the applied field. Therefore, detection of small differences in moment with acceptable signal-to-noise depends both on keeping the sensor in the linear part of its transfer curve (resistance R vs. applied field H) and a high MR ratio. Although single bead magnetic sensors have been reported [13-18], such sensors have been designed to provide information on the number and distribution of beads over the sensor, rather than to distinguish the magnetic moments of individual moment-differentiated beads. Furthermore, no magnetic labeling/detection scheme has been suggested which encodes information on magnetic beads, as is needed to implement a large-scale multiplexed magnetic bioassay. Here, we show quantitative detection of individual magnetic beads using digital MR responses from the magnetic reversals of multilayer ferromagnetic rings.

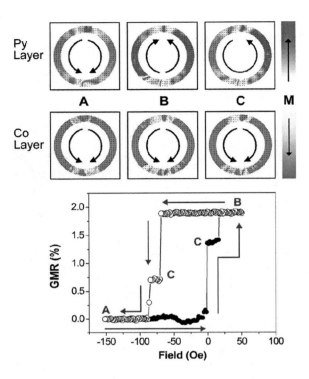

FIGURE 1. Micromagnetic simulations and measured MR response of PSV ring. Micromagnetic simulations using OOMMF [19] show magnetization states of a 2μm PSV ring with 200nm linewidth as the Py layer is cycled from reverse onion (A) to forward onion state (B) and back, as indicated by the arrows. MR measurements on an actual microfabricated ring confirm that the reverse onion state of the Co layer remains largely undisturbed during the cycling, giving rise to minimum (A) and maximum (B) MR levels. The plateaus correspond to intermediate states (C) obtained in the Py layer during the ascending and descending field sweeps. Reprinted from reference [22].

MAGNETIC RING-SHAPED BEAD DETECTORS

The use of ring sensors as sensitive bead detectors was suggested by Miller *et al.* [14], on the basis of the possibility of flux-matching between a ring and the stray field pattern of a magnetic bead magnetized out of plane. In response to an in-plane external magnetic field, microfabricated rings can show highly abrupt, nanosecond-scale switching [20] between their four stable magnetization states – the flux-closed "vortex" states of opposite circulations and the quasi-saturated "onion" states, which contain a head-to-head and a tail-to-tail domain wall [21]. Furthermore, a multilayer pseudo-spin-valve (PSV) ring made by separating two magnetic rings with a thin spacer layer of non-magnetic metal will exhibit giant magnetoresistance (GMR). Figure 1 [22] shows an exemplar minor magnetoresistance loop measured from such a ring with a spacer layer of Cu and soft and hard magnetic layers of $Ni_{80}Fe_{20}$ (permalloy or Py) and Co respectively, and relates the resistance levels to micromagnetic simulations of the ring's magnetization states. Maximum MR is obtained when the Py and Co layers are in opposite onion states, and vice versa; the intermediate MR levels correspond to Py states which resemble the vortex but contain a complex pattern of small reverse domains, correlated by magnetostatic interactions to domains in the Co layer. Variations in this fine magnetization structure account for the difference in the MR levels between the intermediate states obtained on the ascending and descending field sweeps. Importantly for device applications, both circular and elliptical PSV rings have been shown to display abrupt transitions and high signal-to-noise ratio [23, 24], with excellent reproducibility of the switching fields [25] between stable states.

Recently, we showed that this abrupt and highly predictable switching between states allows multilayer rings to be used for bead detection in a way previously unexplored by MR biosensor applications [22]. Digital bead detection cycles the Py layer of the ring between its two onion states A and B using an external field, whilst the Co layer remains in the reverse onion state; in the absence of any beads, the MR output of the ring therefore appears as a train of rectangular pulses with amplitude equal to the maximum MR (binary 1). By tuning the amplitude of the applied field such that it is only just large enough to switch the soft layer when the bead is absent, the stray field from the bead can cancel a fraction of the applied field and prevent the switching of the ring. In this case, the MR output from the ring gives a null signal (binary 0). Digitally detecting beads in this manner ensures a high signal-to-noise ratio, due to use of the full MR response and also because the detection is a differential measurement of the relative positions of one or both of the Py transition fields.

MULTIPLEXED FLUORO-MAGNETIC ASSAY

Figure 2 describes a proposed hybrid fluoro-magnetic method by which a multiplexed bead-based bioassay can be performed using digital sensors. A library of probes labeled with UV-sensitive fluorophores is prepared and immobilized on populations of magnetic beads of different magnetic moments, with only one kind of probe per bead type. The targets are similarly labeled with fluorophores which are sensitive to the emission from the probe fluorophores, but not to UV. Interrogating the sample with UV indicates how well the beads have taken up the probes. After addition of the targets, Fluorescence Resonance Energy Transfer (FRET) [26] between the probe and target fluorophores provides confirmation of hybridization; this effect depends very strongly on the distance between fluorophores and thus is an excellent selector for even small DNA hybridization mismatches. The beads are then passed via microfluidic channels to the magnetic multilayer ring-shaped sensor, which provides a quantitative measurement of their magnetic moment. By comparing the moments of the measured beads which changed color to a bead-sequence look-up table, the probes which successfully hybridized to the target can be identified. The biochip can then be washed and a new set of probes loaded, indexed by a fresh database. This scheme allows the probe sequences to be as long or short as required for the application.

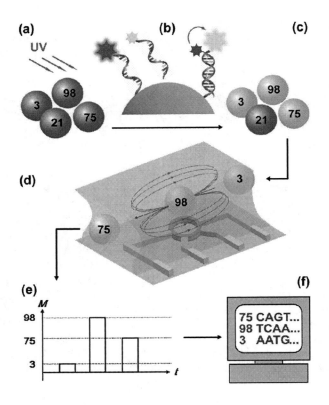

FIGURE 2. Principle of multiplexed magnetic bioassay. An ensemble of beads of different magnetic moments, functionalized with probe DNA and UV-sensitive fluorophores is illuminated with UV to confirm functionalization (a). On mixing the beads with sample solution containing fluorescently labeled analyte DNA (b), FRET confirms and quantifies hybridization for each bead (c); beads which do not change color identify analytes not present in the sample. Microfluidic channels then pass beads to a digital ring sensor, which measures their magnetic moment (d). Biodetection is achieved by comparing the moments of the successfully hybridized beads to a bead-probe look-up database (e). The magnetic biochip can then be washed and a new set of probes loaded, indexed by a fresh table.

Although the biochemistry involved in such an assay is well characterized, several issues must be addressed concerning the digital detection scheme outlined previously. This demands that the shifts of the ring's switching fields must be both well determined and reproducible. It must be possible to determine the magnitude of the field shift with a smaller error than is generated by the distribution of moments of identical beads. This requires a sensor with an abrupt response, because at the lower signal-to-noise levels obtained as the size of the beads decreases, the exact magnitude of the shift becomes increasingly difficult to determine if the MR response is too broad. Also, any feature in the transfer curve used as a yardstick to measure the shift needs to reproducible between beads of nominally identical moments. The problem of reproducibility also becomes more serious as the magnitude of the shift decreases, which places a lower limit on the sizes/moments of beads that can be detected. Previous measurements have shown that the switching of the ring from one state to another is completed within 1 Oe, and that these transitions occur with excellent reproducibility [23, 24, 25]. However, successful digital detection also depends on the sensitivity of the shift of the ring's switching field to the magnetic moment of the beads. This will determine how many beads a sensor can distinguish, and what kind of beads can be used in a moment-encoded assay. Finally, as the moment is determined by the absolute amount of magnetic material in the bead, the manufacturing process for batches of beads intended to encode members of the library cannot permit too large a distribution in the moments of nominally identical beads.

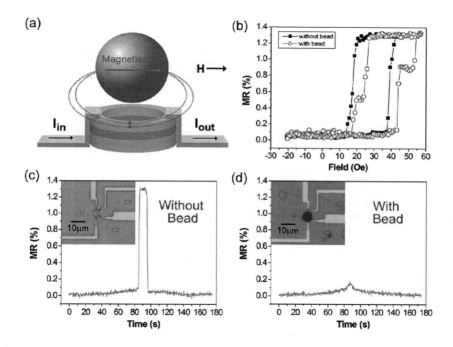

FIGURE 3. Results of bead detection. (a) Schematic showing the induced dipole field from a single magnetic bead partially cancelling the externally applied field over the ring. (b) Minor hysteresis loops of the 4 μm PSV ring taken in both presence (open circles) and absence (solid squares) of a single M-450 Dynabead® after saturating the ring at –1500 Oe. The intermediate state in the Py layer is stabilized in the presence of the bead, and is most likely due to a slight misalignment of the centers of the bead and the ring, which cancels the field to a greater degree at one side of the ring. (c), (d) MR vs. time for a critically balanced loop in the absence and presence (respectively) of an M-450 Dynabead® positioned by a bar magnet over the centre of the ring; optical micrographs are shown inset. The null signal obtained for this field range identifies one M-450 Dynabead®. Adapted from reference [22].

MEASUREMENT OF SINGLE MAGNETIC BEADS

Figure 3 summarizes the results of our initial proof of concept measurements to demonstrate the feasibility of digital bead detection with PSV rings, using 4.5µm diameter M-450 Dynabeads® (Invitrogen Corp.) [27]. The fabrication of the ring sensors and the MR measurement setup are described in a previous paper [22]. The Dynabeads®, which are composed of maghemite nanoparticles encapsulated in monodisperse polymer beads, were mixed with a 1:100 solution of deionised water and Triton®-X 100 (Union Carbide, Inc.) surfactant and deposited on the chip with a hypodermic needle. A single Dynabead® was then positioned with a bar magnet over a PSV ring sensor as shown schematically in Figure 3(a). Figure 3(b) shows minor MR loops taken before (solid squares) and after (open circles) the bead was positioned, clearly showing shifts of the Py layer transitions due to the dipole field generated from the bead. The shifts in the reverse onion – intermediate transition (from 38 Oe to 43 Oe) and the intermediate – reverse onion transition (from 16 to 18 Oe) represent a cancellation of $(13 \pm 0.5)\%$, which corresponds closely to the relative field reduction of 11% predicted by the simple model given in Equation 1, using data from Fonnum et al. [27].

We then demonstrated digital detection of beads by cycling the field between -300 Oe and +41 Oe. This field value was chosen to lie between the switching fields of the Py reverse onion – intermediate transition measured with the bead (the upper bound) and without the bead (the lower bound). Figures 3(c) and 3(d) display the change in the MR signal with time as the field is cycled in the absence and presence of the centrally-placed Dynabead®, respectively. The results clearly show that a large pulse equal to the full MR response is measured in the absence of the bead, and therefore that the Py layer of the ring has switched into the opposite onion state. In contrast, when the bead is present the signal is nulled, indicating that the bead has prevented the switching of the ring. The suppression of switching in the field range represented by this critically balanced loop provides a measure of the magnetic moment, identifying this particular bead. The timescale (180s) reflects the speed at which the DC field was stepped by the magnet power supply; the nanosecond-scale switching of the ring would permit faster bead detection using a high-frequency AC field.

For comparison, bead detection experiments were also done on an elliptical ring co-fabricated on the same sample, which had a long axis of 4µm, a short axis of 2µm, and a linewidth of 219nm. A single Dynabead® was centrally positioned over the ring as previously described and as shown in Figures 4(a) and 4(b). MR measurements and simulations performed by Castaño et al. [24] on similar devices have found that elliptical pseudo-spin-valve rings cycled to and from saturation can exhibit enhanced vortex stability in the Co reversals compared to circular rings. However, the "vortex" state in the Py layer is likely to contain residual domain walls, meaning that the linewidth and layer thicknesses in elliptical rings are critical in determining the stabilities of intermediate Py layer states (which are very small in this sample). This is beneficial for digital bead detection, as the lack of intermediate states increases the abruptness of transitions. Also, the fact that the bead is size-matched to the long

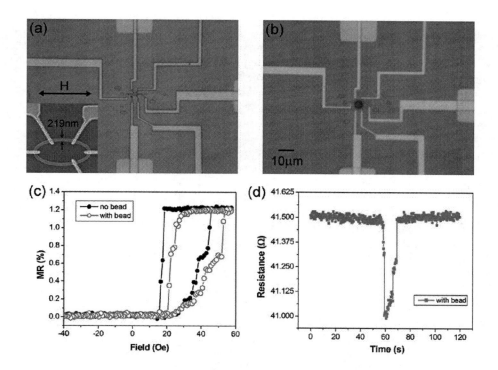

FIGURE 4. (a), (b) A single 4.5μm M-450 Dynabead® is positioned over the elliptical PSV ring (4μm long axis, 2μm short axis, 219nm linewidth). Inset shows scanning electron micrograph of the ring, indicating the linewidth and the applied field direction. (c) Minor loop of Py layer of 4μm elliptical PSV ring, showing transitions in absence (closed circles) and presence (open circles) of the bead. (d) Resistance vs. time response of the elliptical ring for a critically balanced field sweep. An abrupt drop in the resistance occurs in the presence of a single Dynabead®. The ring was supplied with a constant current of 40μA.

axis of the ellipse means that the sides of the ring are much closer to the projected centre of the bead than is assumed for the circular sensor of Equation 1. The cancellation effect induced by the bead that the ring experiences is correspondingly greater; although the switching fields of the Py layer are similar in magnitude to those in the circular ring, the shift is roughly doubled. The reverse onion - onion transition is delayed from 45 to 53 Oe, and the highly abrupt onion - reverse onion transition from 15 to 20 Oe, as shown in Figure 4(c). Measurements taken before and after the bead was positioned are denoted by solid circles and open circles, respectively.

It can immediately be seen that the reversal process in the ellipse is phenomenologically different to that in the circular ring. The Py reverse onion – onion transition seen in the rising branch of the MR loop of Figure 4(c) proceeds via a number of metastable states, which are suspected from simulations to represent multidomain configurations [24], whereas the transition in the falling branch is virtually single-step. Again, vortex-like intermediate states are suppressed in the absence of the bead and stabilized by its presence, though not to such an extent as in the circular case. The characters of the reversals mean that only the onion – reverse onion transition is suitable for use in digital bead detection, as the other is not abrupt enough to gain the full MR response of the ring. A critically balanced loop was therefore defined by saturating the ring at +1500 Oe and cycling the field from -60 Oe to -19 Oe and back. In contrast to the procedure previously employed for the circular ring, this protocol saturates the elliptical ring at positive field, and then cycles the Py layer at negative field in a range which ends just before the nominal transition out of the forward onion state. This critically balanced loop gives a signal only when the bead is present, as would be desirable in a real device. Unfortunately, the comparison between the responses of the elliptical ring in the presence and absence of the bead cannot be shown, as the device failed whilst the first critically balanced loop without the bead was being taken. However, the loop taken with the bead, using a sense current of 40μA (Figure 4(d)), shows a resistance drop of 0.5Ω, equivalent to the ring's full MR response. This method of critically balancing the ring therefore also achieves successful single-bead detection, but now in the manner required for an actual biosensor.

CONCLUSIONS

We have shown that a PSV ring can detect a single size-matched bead and measure its moment in a digital manner, using a simple model. The sensitivity of the movement of domain walls in the ring to the bead's magnetostatic field means that the ring could be used in the conventional binding assay to detect high-moment beads with diameters much less than that of the ring, for example by functionalizing a small area of the ring annulus through an overlaid lithographic template [28]. However, a much higher throughput can be obtained if all biochemical reactions take place on the bead, and the sensor detects the beads in flow; the ensemble of beads can then be characterized by their measured magnetic moment via the magnitude of the switching field cancellation they induce, normalized to that induced by the bead with the smallest moment. This scheme depends on fine control of the position of beads in the microfluidic channel relative to the ring sensor, using e.g. dielectrophoretic [29] and/or magnetic [30] field gradients, and is equally dependent on the moment distribution of nominally identical beads. We are currently developing an integrated microfluidic device to carry out a multiplexed bead-based bioassay using a panel of antigens, labeled by an ensemble of moment-encoded magnetic beads. Magnetic encoding integrated with libraries of biologically relevant compounds should provide a significant increase in accuracy, throughput, and multiplexing capacity in performing bioassays.

ACKNOWLEDGMENTS

The authors would like to thank R.B. Balsod and S. Sigurdsson of the Cavendish Laboratory Workshops for design and construction of the sample holder used to perform the measurements, and Professor H. I. Smith (MIT) for use of nanofabrication facilities by which the ring sensors were manufactured. Financial support for the research is gratefully acknowledged from the EPSRC under the Basic Technology Research Programme "4G Project" (J.L., T.J.H, and J.A.C.B.), the Singapore-MIT Alliance (F.J.C., I.A.C., and C.A.R), and a Marie Curie Fellowship (D.M.).

REFERENCES

1. M. Han, X. Gao, J.Z. Su and S. Nie, *Nat. Biotechnol.* **19**, 631-635 (2001).
2. K. Braeckmans, S.C. de Smedt, C. Roelant, M. Leblans, R. Pauwels and J. Demeester, *Nat. Mater.* **2**, 169-173 (2003).
3. S.P. Mulvaney, H.M. Mattoussi and L.J. Whitman, *Biotechniques* **36**, 602-609 (2004).
4. P.S. Eastman, W. Ruan, M. Doctolero, R. Nuttall, G.D. Feo, J.S. Park, J.S.F. Chu, P. Cooke, J.W. Gray, S. Li and F.F. Chen, *Nano Lett.* **6**, 1059-1064 (2006).
5. Y.W.C. Cao, R.C. Jin and C.A. Mirkin, *Science* **297**, 1536-1540 (2002).
6. S.R. Nicewarner-Peña, R.G. Freeman, B.D. Reiss, L. He, D.J. Peña, I.D. Walton, R. Cromer, C.D. Keating and M.J. Natan, *Science* **294**, 137-141 (2001).
7. D.C. Pregibon, M.Toner and P.S. Doyle, *Science* **315**, 1393-1396 (2007).
8. G.S. Galitonov, S.W. Birtwell, N.I. Zheludev and H.Morgan, *Opt. Express* **14**, 1382-1387 (2006).
9. D. Axelrod, T.P. Burghardt and N.L. Thompson, *Annu. Rev. Biophys. Bioeng.* **13**, 247-268 (1984).
10. V. Joshi, G. Li, S.X. Wang and S. Sun, *IEEE Trans. Magn.* **40**, 3012-3014 (2004).
11. D.R. Baselt, G.U. Lee, M. Natesan, S.W. Metzger, P.E. Sheehan and R. J. Colton, *Biosens. Bioelect.* **13**, 731-739 (1998).
12. J.D. Jackson, *Classical Electrodynamics* (3rd ed.), New York: J. Wiley & Sons, 1998, pp. 186-200.
13. L. Ejsing, M. Hansen, A. Menon, H.A. Ferreira, D.L. Graham and P.P. Freitas, *J. Magn. Magn. Mater.* **293**, 677-684 (2005).
14. M.M. Miller, G.A. Prinz, S.F. Cheng and S. Bounnak, *Appl. Phys. Lett.* **81**, 2211-2213 (2002).
15. G. Li, V. Joshi, R. White, S.X. Wang, J. Kemp, C. Webb, R. Davis and S. Sun, *J. Appl. Phys.* **93**, 7557-7559 (2003).
16. H.A. Ferreira, D.L. Graham, P.P. Freitas and J.M.S. Cabral, *J. Appl. Phys.* **93**, 7281-7286 (2003).
17. W. Shen, X. Liu, D. Mazumdar and G. Xiao, *Appl. Phys. Lett.* **86**, 253901 (2005).
18. S. Mihajlović, P. Xiong, S. von Molnár, K. Ohtani, H. Ohno, M. Field and G. Sullivan, *Appl. Phys. Lett.* **87**, 112502 (2005).
19. M.J. Donahue and D.G. Porter, *Interagency Report* **NISTIR 6376** (http://math.nist.gov/oommf) (1999).
20. M. Kläui, C.A.F. Vaz, L. Lopez-Diaz and J.A.C. Bland, *J. Phys.: Condens. Matter* **15**, R985-1023 (2003).
21. J. Rothman, M. Kläui, L. Lopez-Diaz, C.A.F. Vaz, A. Bleloch, J.A.C. Bland, Z. Cui and R. Speaks, *Phys. Rev. Lett.* **86**, 1098-1101 (2001).
22. J. Llandro, T.J. Hayward, D. Morecroft, J.A.C. Bland, F.J. Castaño, I.A. Colin and C.A. Ross, *Appl. Phys. Lett.* **91**, 203904 (2007).
23. T.J. Hayward, J. Llandro, R.B. Balsod, J.A.C. Bland, F.J. Castaño, D. Morecroft and C.A. Ross, *Phys. Rev. B* **74**, 134405 (2006).
24. F.J. Castaño, D. Morecroft and C.A. Ross, *Phys. Rev. B* **74**, 224401 (2006).
25. T.J. Hayward, J. Llandro, F.D.O. Schackert, D. Morecroft, R.B. Balsod, J.A.C. Bland, F.J. Castaño, and C.A. Ross, *J. Phys. D: Appl. Phys.* **40**, 1273-1279 (2007).
26. J.R. Lakowicz, *Principles of Fluorescence Spectroscopy* (2nd ed.), New York: Springer Science-Business, 1999.
27. G. Fonnum, C. Johansson, A. Molteburg, S. Mørup and E. Aksnes, *J. Magn. Magn. Mater.* **293**, 41-47 (2005).
28. G. Li, S. Sun, R.J. Wilson, R.L. White, N. Pourmand and S. X. Wang, *Sens. Act. A* **126**, 98-106 (2006).
29. D. Holmes, H. Morgan and N.G. Green, *Biosens. Bioelectron.* **21**, 1621-1630 (2006).
30. Z. Jiang, J. Llandro, T. Mitrelias and J. A. C. Bland, *J. Appl. Phys.* **99**, 08S105 (2006).

Advanced Magnetoresistance Sensing of Rotation Rate for Biomedical Applications

Marioara Avram[a], Marius Volmer[b], and Andrei Avram[a]

[a]National Institute for Research and Development in Microtechnologies, 126A, Erou Iancu Nicolae street, 077190, Bucharest, ROMANIA
[b]Transilvania University, 29 Eroilor Boulevard, 500036, Brasov, ROMANIA

Abstract. We propose to build a non-Newtonian fluids viscosimeter, in order to measure the viscosity of biological fluids such as blood. The system is based on a rotating microgear wheel and a magnetoresistive sensor with a non-contacting transduction mechanism to transform the rotor rotation rate into an electrical signal. As the rotor turns, the field from this microscopic magnet will modulate the resistance of a bar of a low coercitivity material such as Permalloy, with an in-plane uniaxial magnetization, placed nearby, close to the rotor flanges. The change in resistivity provides an electrical signal with frequency proportional to the rotation rate, and hence the fluid velocity. The rotor is fabricated from polysilicon and coated with a soft magnetic material. The magnetoresistive sensor is formed of two Wheatstone bridges orientated on the X and Y axes. As the microgear wheel rotates, a tooth passing by the sensing GMR of the Wheatstone bridge changes the magnetic field, thus enabling us to measure the velocity of the gear wheel. The gear wheel has the outer diameter of 200 μm and is obtained by using the cut and refill technique. The basis for fabrication of movable parts is the use of sacrificial layers that act both as spacers and also to keep the parts attached to the silicon wafer during fabrication.

Keywords: Microfluidic viscosimeter, Biological fluids, Micro gear wheels.

PACS: 47. 60. Dx, 85. 70. Ay, 85. 75 -d

INTRODUCTION

Current methods used for measuring blood viscosity are out-of-date. They have many disadvantages, such as: long measurement time, many disturbance factors may appear during measurement, the blood sample may be damaged during the process, etc. At the moment, rapid, accurate and repeatable measurements are not possible. A new method to measure blood viscosity is proposed in this paper, which uses a needle instead of capillary and can measure blood viscosity while collecting blood. Owing to the advantages of small amount of blood sample, rapidity and repeatability, the new method is of value for clinical application. The paper presents a magnetoresistive sensor used to detect the rotation rate of a micro-dynamic gear wheel system. The sensor itself is a double Wheatstone bridge built with giant magnetoresistance (GMR) arranged on two perpendicular axes. Magnetic material, such as Pemalloy, deposited on the micro-dynamic system disturbs the magnetic field of the sensing GMR's, thus

CP1025, *Biomagnetism and Magnetic Biosystems Based on Molecular Recognition Processes*
edited by J. A. C. Bland and A. Ionescu
© 2008 American Institute of Physics 978-0-7354-0547-9/08/$23.00

creating an electrical signal. The electrical potential from the output of the double Wheatstone bridge is an indicator of the microgear wheel's velocity.

THEORETICAL ASPECTS

Human blood varies substantially from person to person, with state of health and other factors. Typical values of the density of blood plasma and blood cells are $1025 kg/m^3$ and $1125 kg/m^3$, respectively. The viscosity of blood at 37°C is about 2.7 centipoises, much greater than the viscosity of water at the same temperature (about 0.7 centipoises). The flow of blood in the human circulatory system, which consists of nearly circular cross section blood vessels, can be described quite well using Poiseuille's equation:

$$\frac{V}{t} = \frac{\pi(P_1 - P_2)R^4}{8L\mu}.$$
(1)

In medicine, when referring to the viscosity of blood, we talk about kinematic viscosity, which is denoted with v and has the SI unit ($mm^2 \cdot s^{-1}$). Rotational viscosimeters, such as our micro-gearwheel viscosimeter, measure the dynamic viscosity, which is denoted with η and has the SI unit (Pa·s). Usually, in literature, the dynamic viscosity is measured in cP (centipoise, 1cP = 1mPa·s).

Let us consider the movement of fluid between a gear wheel and a cylinder, the gear wheel having the extreme radius "a" and the cylinder having the radius "b". Let ω be the angular velocity of the fluid at distance r from the axis of rotation, and let the shearing stress and rate of shear at this radius be F_r and D_r, respectively.

$$F_r = \frac{M}{2\pi r^2 h} \quad \text{and}$$
$$D_r = r\frac{d\omega}{dr}$$
(2)

$$M = \frac{4\pi a^2 b^2 h \eta \omega}{b^2 - a^2}.$$
(3)

For a Newtonian fluid, $F_r = \eta D_r$, and

$$\frac{M}{2\pi r^2 h} = \eta r \frac{d\omega}{dr}.$$
(4)

There are many biological fluids, particularly those in the form of more concentrated suspensions and emulsions that are non-Newtonian. These include the Bingham fluids, for which the rate of shear is zero if the shearing stress is less than or equal to a yield stress, f, and is otherwise directly proportional to the shearing stress in excess of the yield stress [1]

$$F = f + \frac{1}{\mu} D .$$ (5)

The constant μ is the mobility.

A family of materials of ever growing importance in the field of synthetic polymers is described as visco-elastic because they combine some of the properties of a viscous fluid with some of the properties of an elastic solid. Such materials require specialized techniques to elucidate the relationship between stress, rate of shear and amount of shear. A detailed consideration of visco-elastic biological materials is outside the scope of this paper.

The term viscosity of non-Newtonian fluids, η_B, is defined as:

$$\eta_B = \frac{1}{\mu} + \frac{f}{D} .$$ (6)

The rotational viscometer is consequently widely used for the study of the flow properties of non-Newtonian fluids. The rotational viscometer comprises of two parts, separated by the fluid under test, which are able to rotate relative one to another about a common axis of symmetry. As a part rotates, the other tends to be dragged around with it due to the torque (M) transmitted by the test fluid.

If the fluid obeys Bingham's law, equation (5) can be calculated:

(i) if $M < 2\pi a^2 hf$, the shearing stress throughout the fluid is less than the yield stress and no shear occurs;

(ii) if $2\pi a^2 hf < M < 2\pi b^2 hf$, the shearing stress exceeds the yield stress only in the region between the rotor and a critical radius r_c, given by

$$r_c = \sqrt{\frac{M}{2\pi hf}} ;$$ (7)

(iii) if $M > 2\pi b^2 hf$, the shearing stress exceeds the yield value at all points. For this case, integration leads to the Reiner – Riwlin equation:

$$\omega = \frac{M}{2\pi h \cdot \eta_B} \left(\frac{1}{a^2} - \frac{1}{b^2} \right) - \frac{f}{\eta_B} \ln \frac{b}{a} ,$$ (8)

where η_B is the viscosity of a non-Newtonian Bingham fluid.

DEVICE DESIGN

The sensor uses two Wheatstone bridges consisting of GMR resistors to relay the rotation rate information from the microdynamic gearwheel system. This system is a non-contacting transduction mechanism. It has the advantage of reduced frictional drag on the moving parts and it is expected to show a high reliability [2]. We propose an ensemble of a magnetoresistive sensor with a non-contacting transduction mechanism to transform the rotor rotation rate into an electrical signal. The two Wheatstone bridge sensors are arranged along the X and Y directions for an improved sensing capability. The sensitive structure uses the two sensors to measure both the X component of the field and the Y component of the field at the same point (Fig.1). Both sensors are full Wheatstone bridge configurations with four active GMR resistors, placed in the middle of the sensitive structure, and four shielded reference GMR resistors, placed near the edges.

The sensing resistors are located between flux concentrators in order to direct the magnetic field on them and have large magnetic field variations to generate an electric signal. The reference resistors are shielded in order to protect them from any influencing magnetic field. All resistors were fabricated from thin multilayer (12 thin metallic layers alternating magnetic – Permalloy (4 nm) and non-magnetic (2 nm) – Cu/Au5%/Ag5%) deposited in a sputtering system at 200 W and 30 mTorr [3, 4]. The sensitive direction of the resistors is perpendicular to the deposited layers because the resistors are made up of multiple narrow stripes of alternating magnetic and non-magnetic materials. The sensor will have the largest output signal when the magnetic field of interest is parallel to the flux concentrator axis.

The differential sensors provide an output signal by sensing the gradient of the magnetic field across the sensor. When a magnetic field approaches the sensor array (Fig.1) from the right (down), the two resistive sensor elements will decrease in resistance. When an external field is applied, the exposed resistors decrease in

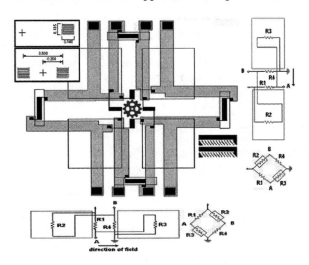

FIGURE 1. The two-axis independent Wheatstone bridges sensor layout.

electrical resistance while the other resistor pair remain unchanged, causing an imbalance condition in the two bridges, providing a signal output from the bridges terminals. A permanent magnet is required to generate a magnetic bias field. The differential magnetic sensor will then be used to detect magnetic gradient field of the permanent magnet as the gear tooth passes by in close proximity [5]. The GMR Wheatstone Bridge sensor array was combined with an operational amplifier (gains of several thousand are easily achieved), high-pass and low-pass filters formed from passive components will be incorporated in the circuit to limit noise and avoid the saturation of the amplifiers by any offset or by a magnetic signal such Earth's field.

The central gearwheel is fabricated using the sacrificial layers technique and the cut-and-refill technique. Its outer diameter is 200 μm and it has 24 teeth.

BIOANALYTICAL DEVICE FABRICATION

Microchannels Fabrication

Microchannels are often used to mimic capillaries, which allow the observation of cells squeezing through the narrow passages of the vascular system. Meanwhile, single cell analysis and measurement of transit time, total flow rate of cell suspensions and cell component aggregation can be studied using microchannels. Photolithography techniques were also used to create hydrophilic and hydrophobic areas inside microchannels [6] by patterning the interior of the protein-coated channels. The ultraviolet light actually cleaves exposed protein molecules, changing the proteins from hydrophobic to hydrophilic. When flow was then activated through the channels, the fluid wetted only the patterned hydrophilic areas. For our device the channel length was chosen to be 150 μm to allow enough length for cell travel and maintain a relatively low channel resistance. The input and output fluid ports were etched completely through the wafer thickness creating holes in the chip. The ports are $1mm^2$ area for connection purpose. The ports are centered arbitrarily inside the reservoirs. The fluid reservoirs are simply a storage area for fluid before it is pushed into the channels and after it leaves the channels. In addition, a scale bar was drawn above and below each of the channel device. The bar consists of 3 μm wide rectangles that are spaced 3 μm apart. This allows measurement of cell length and travel distance.

Gearwheels System Fabrication

The fabrication processes of the gear-wheels system combines the undercut and refill techniques with pin-joints bearing permitting the fabrication of bushings that can be used to elevate the rotor away from the silicon surface. The basis for fabrication of movable parts is the use of SiO_2 sacrificial layers that act both as spacers and also keep the parts attached to the silicon wafer during fabrication [7]. On a 4" Corning 7740 wafer a 4 μm-thick amorphous Si low stress was deposited followed by a 3 μm-thick SiO_2 layer and a 3 μm-thick polysilicon deposited in a horizontal furnace. The polysilicon layer is anisotropic patterned using a DRIE process through a photoresist mask (the body and the teeth of the wheels are defined during this step). The etching process stops on the SiO_2 layer. Using a second photoresist mask the hole of the wheels was etched using the same DRIE process. The SiO_2 layer is etched through the photoresist-polysilicon mask. The isotropic wet etching was preformed with an underetching of the polysilicon layer. This underetching generates a space that will be filled with SiO_2 and polysilicon in the next steps (required for the flange fabrication). After removing the photoresist a new 1-μm LTO layer is deposited in furnace (for a conformal deposition). The thickness of this layer set the tolerance of the adjustment. Through a photoresist mask (deposited by spray coating for a better coverage) the oxide layer deposited on the bottom of hole is removed by wet etching. The step is absolutely necessary in order to create the mechanical contact of the flange on the silicon substrate, below the wheel. A second polysilicon layer (3 μm-thick) is deposited in a horizontal furnace. The conformal deposition assures the filing of the gap under the wheel generated during the previous steps. The second layer of polysilicon is pattern (using DRIE) through a photoresist mask in order to define the shape of the flange below the wheel. In the last step, the SiO_2 molding layers are removed in HF 49% and the gearwheel structure is released. In order to avoid the bending of the polysilicon layer the stress in this layer was adjust by annealing – method suggested by Chen et al in [8]. The measurement of the stress value was performed on test wafer using a KLA-Tencor stress measurement system. The sticking effect of the polysilicon layer on silicon surface was avoided by a methanol rinse after dissolving the latter sacrificial layer follow by drying process with CO_2 in a Critical Point Dryer (Baltec) [7].

MEASUREMENT SET-UP

The measuring system is an electro-mechanic momentum changing device, based on the relative rotation of the gear wheel coupled with the rally axis. A dynamometer, tied in a bridge with a potentiometer, measures the relative rotation such that the signal obtained is proportional with the momentum which acts on the indicator. Switching to different positions on the dynamometer allows us to change the torque [9].

For each experimental point, we have calculated the geometric mean shearing stress F_g using equation (9) and we have ploted the values of angular velocity, ω, against $\ln(M)$.

$$F_g = \sqrt{F_1 \cdot F_2} = \frac{M}{2\pi \cdot a \cdot b \cdot h}. \tag{9}$$

The true shear rate coresponding to F_g is then obtain using Krieger and Elrod formula:

$$D = \frac{\omega}{k} - \frac{k}{6}\frac{d^2\omega}{d(\ln M)^2} + \frac{7k^3}{360}\frac{d^4\omega}{d(\ln M)^4} \quad \text{and} \quad k = \ln\left(\frac{b}{a}\right). \tag{10}$$

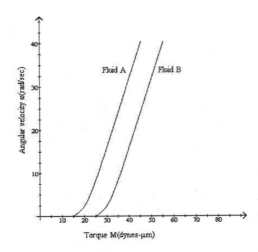

FIGURE 2. Flow curves for two non-Newtonian colloidal organic fluids; fluid B has a higher slurry concentration then the fluid A.

A concentrated slurry in a Newtonian liquid may exhibit Bingham properties. The flow curves for two slurry fluids, fluid B more slurry concentrated than fluid A, are shown in Fig. 2. The non-linear flow curves observed for the two fluids can be explained by interaction between particles, interaction with the continuous phase, and particle deformation. Bingham's law of plastic flow gives us

$$\eta_B = \frac{F_g - f}{-\dfrac{dv}{dr}}. \tag{11}$$

The rate of shear at the rotor is determined for the straight portion of the flow curve. This is done by solving equation (8) for η_B from equation (11). The shear rate is plotted against the shear stress at the rotor from the measured torque. The slope of this plot is η_B, by definition.

CONCLUSIONS

The purpose of this research is to design and fabricate a microfluidic biodynamic system to measure the velocity and viscosity of biological non-Newtonian fluids (e.g. blood in coronary arteries, cerebrospinal fluid). The flow microsensor was designed and fabricated from polysilicon thin films through silicon surface micromachining and sacrificial layers technique. The microfluidic sensor consists of a driving gearwheel with external diameters of 200 μm and 3 μm thickness. The wheel needs clearances to guarantee the motion and bearings for centering. The microsensor design has a flow channel to guide the fluid across the rotor, causing a velocity gradient which is necessary to make the gear wheels system turn. This micro device enjoys the advantage of being compatible with silicon IC fabrication technology. The device was

192

designed for asymmetrical forces to act on the rotor. Fluidic parameters such as dynamic contact angle or capillarity were taken into account for optimizing the microfluidic system performance. The magnetoresistive sensor is formed of two Wheatstone bridges oriented on the X and Y axes. As the micro gear wheel rotates, when a tooth passes by the sensing GMR of the Wheatstone bridge the magnetic field changes, thus enabling to measure the velocity of the gear wheel.

ACKNOWLEDGEMENT

This project is funded by the Romanian Minister of Education and Research, CEEX Program, Project Nr. 27/10.10.2005.

REFERENCES

1. M.A. Avram, M. Avram, C. Iliescu and A. Bragaru, "Flow of non-Newtonian fluids" in *CAS 2006 Proceedings* (an IEEE publication) **2**, 2006, pp. 433-436.
2. C. Reig, D. Ramirez, S. Cardoso and P.P. Freitas, *Proceedings Eurosensors XIX*, 2005, pp Wc5.
3. M. Avram, "Deposition experiments of thin metallic multilayers with magnetoresistive properties", *Journal of Optoelectronics and Advanced Materials* **6**, 987-990 (2004).
4. K. Reichelt and X. Jiang, *Vacuum* **42**, no.1-2, 171-173 (1991).
5. B.R. Jackson and T. Pitman, U.S. Patent No. 6,345,224 (8 July 2004).
6. L.D. Landau and E.M. Lifshitz, *Fluid Mechanics (Course of Theoretical Physics Volume* 6), Pergamon Press, London, 1959.
7. D.P. Poenar, C. Iliescu, M. Carp, A.J. Pang and K.J. Leck, "Glass-based microfluidic device fabricated by parylene wafer-to-wafer bonding for impedance spectroscopy", *Sensors and Actuators A* **139**, 404-411 (2007).
8. L. Chen, J. Miao, L. Guo and R. Lin, *Surface and Coatings Technology* **141/1**, 96-102 (2001).
9. M. Avram, M.A. Avram and C. Iliescu, "Biodynamical analysis microfluidic system", *Microelectronic Engineering* **83**, issue 4-9, 1688-1691 (2006).

AUTHOR INDEX

A

Andò, B., 139
Ascia, A., 139
Avram, A., 186
Avram, M., 125, 186

B

Baglio, S., 139
Barandiarán, J. M., 131
Barbagallo, M., 52
Barnes, C. H. W., 111
Bethge, O., 28
Bland, J. A. C., 34, 52, 60, 74, 111, 176
Brückl, H., 28
Bulsara, A. R., 139

C

Cammarata, R. C., 44
Cardoso, F. A., 150
Cardoso, S., 150
Castaño, F. J., 176
Chien, C. L., 44
Colin, I. A., 176

D

Darling, D., 20
Darton, N. J., 20

E

Eggeling, M., 28

F

Fal-Miyar, V., 131
Fan, D. L., 44
Farzaneh, F., 20
Ferreira, R., 150

Fonesca, L. P., 150
Freitas, P. P., 150
Frey, N. A., 131

G

Gandy, A. P., 82
Germano, J., 150

H

Hallmark, B., 20
Han, X., 20
Hayward, T. J., 74, 111, 176
Heer, R., 28
Hong, B., 74

I

In, V., 139
Ionescu, A., 52, 74

J

Jeong, J.-R., 74

K

Kataeva, N., 28
Kopper, K. P., 60, 74, 111
Kumar, A., 131
Kurlyandskaya, G. V., 131

L

Lew, W. S., 34
Llandro, J., 74, 111, 176
Loureiro, J., 150